PERSONALITY DEVELOPMENT

K. K. Singh

Gaurav Book Centre Pvt. Ltd.
New Delhi

Publisher
Gaurav Book Centre Pvt. Ltd.
4832/24,Prahlad Lane,S-207 Ansari, Road,
Daryaganj, Delhi-110002 Ph.: 43570976, 23278261
Email: gauravbookcentre@gmail.com

Edition: 2015

ISBN : 978-93-83316-06-9

Laser Typesetting
JEE-VEE Graphics Delhi

Price : 950.00

Printed by
Vikas Computers

PREFACE

This book has been designed to meet the requirements of HR professional, Executives who are performing their jobs at Corporate offices and students studying at schools and colleges, the boys preparing for competitive examinations and moreover who are running Group of Companies, Group of Institutions and other educational and financial establishment in the different parts of India or abroad. It would be helpful for all who are dealing with Personality profession as a whole in the HRM Sectors. It will help in making right person for right position.

The present book is a novel attempt to cover a wide range of the problems of Personality Development in the segment of Development of Personality among the people of all kinds of personalities in the human beings of the world. This book faces very busy ideological preparedness in my life. It is in the context of managerial work that we have studied this aspect of English literature, but in this book it has by no means limited to that context. Busy idleness is a disease that affects everybody and pervades every aspect of life. Active non-action is the word that I would like to explain here.

In our professional life, we always face problems in the shape of daily problem solving activities. Problems require a big picture perspective—which means reflection, systematic planning, creative thinking and time. Operational activities squeeze important problems out. Daily routines, superficial behavior, poorly prioritized unfocused tasks leech managers capacities- making unproductive business.

The problem is not a lack of knowledge or resources. The real problem is that they simply do not do those things. They spend their time spinning their wheels, attending meetings and responding to every little query and problem.

Management is "the art of getting things done." Managers must act themselves and mobilize collective action on the part of others. The gap between knowledge and action, stretches wide and few managers seem able to cross it. The kind of behavior that exhibited active non-action as it is called pervasive corporate "knowing-doing gap." Managers always complain about the problem of active non-action but have not fully understood the underlying dynamics.

When driven by deep personal commitment to the goal that cuts out distractions and overcomes difficulties. The purposeful action is not a quick shuffle or mere flirtation with ideas. It is action taking to produce certain results with undivided resolve. Although external issues in an organization can make this kind of action taking difficulties.

We are deeply indebted to all the Websites and Academicians, Researchers, HR professionals and all Managers who participated directly or indirectly in authoring this book, giving us generously not only their time but also their knowledge and perspectives. I have mentioned a few of them in the book, but many others who provided their cooperation a lot go unnamed. I pay my deep acknowledgement from my core of heart to all concerned people to pay their hearty as well as practical help to this book.

And to so many others who have been left unnamed. We are also grateful to the United States of America and United Kingdom of England for their great help from Educational Associations of Management Studies, which provided a management literature, without that nothing was possible. I am also thankful to community of colleagues whose help and advice improved the quality of our thinking and portrayal of this book.

I have described in this book are as follows:

01. What is Personality?
02. Theories of Personality;
03. What is Personality Disorder?
04. Traits Theories;
05. The Measurement of Personality;
06. The six-stages of Modern Career Development;
07. Personality Conflict;

08. Psychological Theories of Personality Development;
09. Self Discipline;
10. How your Emotions Affect your Health;
11. Personality and Masks
12. Dependent Personality Disorder;
13. What makes an Effective Leader?
14. Personality related clinical Terms and Definitions.

This book seeks to serve as a reliable guide and companion to all employers, Business Owners, HR professionals and all those who are involved directly or indirectly with the businesses of different groups. Although efforts are required for a drastic change in the Management studies in India or in abroad, yet it will still take quite some time to get those proposals translated into reality. With more emphasis on the relationships among Owners, HR professional and workmen, the improvement on organization productivity has not been crossed the appropriate level of the present requirement. In this context I would like to enunciate that the Capitalism or hoarding of resources at the center point it will not do the organization`s problems to be resolved in the near future. Resources are meant for all not for counted people in the segment of businesses or with some Temple Estates, so called Gurus and so many types of property hoarders which are becoming gorgeously richer and richer and our poverty stricken people are in dearth of even two loaves of bread a day.

At one place, abundance of lavishness and aristocracy is prevailed and at another, there is nothing to be pulled on their life. This ingratitude must be worn out from the society of India. Now days, few people are busy with the hoarding of the cash and kind from all means. They don`t have any thinking behind it that how the rest of the population will survive even with appropriate needs of the survival.

That is why the need to know the rudimentary rules and rubrics of law for all, as these is necessary. Much has been changing in general laws but nothing we find in shape of progress and prosperity. The basic tenet and underpinning philosophy of general laws is to protect the interests of the people by saving them from exploitation and harassment on one hand and providing enough leverage and stick of good governance to the Government so as to have concrete art of living culture on the other hand. The cordial and congenial relationship can be had only when both the limbs of society have better understanding in the light of the limits of their rights, responsibilities and duties.

As stated ibid, they are something serious rather than contextual but everything is fair in LOVE and WAR. Here love is for the public of India and war is against the cruel people of society who are behaving like blackcaps of India.

Most of the Indian professionals are unable to forget work while on a holiday, working up to three hours instead of enjoying the free time with family and friends. Developments in technology mean that workers are always connected and the temptation to check emails and complete any tasks that follow on from them is easy to succumb to.

It does not matter that only a few in each generation will grasp and achieve the full reality of man`s proper stature—and that the rest will betray it. It is those few that move the world and give life its meaning—and it is these few that have always sought to address. The rest are no concern of mine; it is not me or all others that they will betray: it is their own souls.

Now, it is, I am hoping in very compassionate way that this book will be proved to be in positive manner, boon for the readers.

2013

Krishna Kumar Singh

FOREWORD

Personality Development quintessentially means enhancing and grooming one's outer and inner self to bring about a positive change to your life. Each individual has a distinct persona that can be developed, polished and refined. This process includes boosting one's confidence, improving communication and language speaking abilities, widening ones scope of knowledge, developing certain hobbies or skills, learning fine etiquettes and manners, adding style and grace to the way one looks, talks and walks and overall imbibing oneself with positivity, liveliness and peace.

The whole process of this development takes place over a period of time. Even though there are many crash courses in personality development that are made available to people of all age groups, implementing this to your routine and bringing about a positive change in oneself takes a considerable amount of time. It is not necessary to join a personality development course; one can take a few tips and develop his or her own aura or charm.

- You may have heard this a million times "Think Positive". It works.
- Smile. And smile some more. It adds to your face value and to your personality as well.
- Read a few articles in the newspaper loudly. This will help in communicating fluently.
- Follow table manners and dining etiquettes
- Take good care of your health, dress well, be neat and organized
- Prepare a chart that mentions your strengths and weaknesses. Now concentrate on the latter and find ways to improve upon the same. Do not forget to strengthen your strengths.
- Spend some time alone concentrating on you and yourself alone.

- Practice meditation and yoga. It will help you develop inner peace and harmony that will reflect outside.
- Do not live a monotonous life. Be creative and do something new all the time. Nothing bigger than the joy of creative satisfaction.

Personality development is gaining more and more importance because it enables people to create a good impression about themselves on others; it helps them to build and develop relationships, helps in your career growth and also helps to improve your financial needs.

After all, personality development is nothing but a tool that helps you realize your capabilities and your strengths making you a stronger, a happier and a cheerful person.

Contents

1

WHAT IS PERSONALITY?

Personality is done up of the characteristic patterns of thoughts, feelings and behaviours that make a person unique.

Almost every day we all describe and assess the personalities of the people around us. We realize these daily contemplations on how and why people behave personality psychologists do in the similar manner or not are the matter to be critically analysed.

While we assess personality informally tend to focus more on individuals, instead personality psychologists conduct conceptions of personality that can put into effect to everyone. The works on Personality taken to the development of a number of theories that help explain how or why certain personality traits develop.

Definition of Personality

Personality is a flat out of habits, traits, characteristics of an individual. It is actually a force or power to do things. Some people have more power and some have less e.g. an army officer has the power and knowledge about security issues whereas a professor may have knowledge and powers on educational issues.

Personality – the complex of all the attributes – behavioral, temperamental, emotional and mental – that characterize a unique individual; "their different reactions reflected their very different personalities"; "it is his nature to help others"

Individual, mortal, person, somebody, someone, soul – a human being; "there was too much for one person to do".

Attribute – an abstraction belonging to or characteristic of an entity.

Trait – a distinguishing feature of your personal nature

Identity, personal identity, individuality – the distinct personality of an individual regarded as a persisting entity; "you can lose your identity when you join the army"

Personableness – the complex of attributes that make a person socially attractive

Anal personality, anal retentive personality – (Psychoanalysis) a personality characterized by meticulous neatness and suspicion and reserve; said to be formed in early childhood by fixation during the anal stage of development.

Genital personality – (psychoanalysis) the mature personality which is not dominated by infantile pleasure drives.

Narcissistic personality – personality marked by self-love and self-absorption; unrealistic views about your own qualities and has little regard for others.

Obsessive-compulsive personality – he is personalized by a strong need to repeat certain acts or rituals.

Oral personality – (psychoanalysis) a personality characterized either by generous optimism or aggressive and ambitious selfishness; formed in early childhood by fixation during the oral stage of development.

Character, fibre, fiber – the inherent complex of attributes that determines a person`s moral and ethical actions and reactions; "education has for its object the formation of character" – Herbert Spencer.

Nature – the complex of emotional and intellectual attributes that determine a person`s characteristic actions and reactions; "it is his nature to help others"

Personality – a person of considerable prominence; "she is a Bollywood personality.

Celebrity, famous person – a widely known person; "he was a cricket celebrity.

Another important aspect of personality is intelligence quotient. In a job situation employers are interested to see how quickly employees can pick up instructions, learn more new skills and adapt to the change in demands such skills are rewarded and helpful in developing a successful career for the individuals and increase in organizations profit.

In today's organizations people from different cultures and sub-cultures come to work. If someone comes from a rural or urban back ground they will have different ways of looking at different entities which makes it harder to deal with them. It is difficult to get work done by these people because they will not work if they are dealt with in same monotonous manner. How singular people observe unlike things because of their cultural back ground. In an organization, when individuals are asked to perform a task every one may not be willing to do it depending on how they perceive their orders. Human relations becomes important in such situations if the supervisor can access the situation, mood, and personality of the employees and then give orders and get the work done. The supervisors should appreciate individual diversity and act accordingly.

Personality is basically the sum total of our inner and outer attributes; while outer personality helps us to be accepted, liked and loved, inner traits make long lasting impressions and relationships helping us suceed and grow throughout our lives.

Personality Development is a continuous process which a person at any time develops through conscious efforts towards the attainment of a 'better self'.

The term personality refers basically to the type of person an individual is. The word comes from the Middle English word "personalite", taken from the Old French, and derived from the Latin word "persona" meaning person. The term refers to the description of the behavioural, mental and emotional traits of a human being. It is a term referred to, to quiet an extent in personality psychology; which deals with such things as personality types, personality disorders etc. For a lay man the personality of a person can be defined as pleasing, short tempered, arrogant etc based on their obvious deportment.

Personality can also be used in an abstract sense to refer to the characteristics or feel of a book, a car etc.

Totality of an individual's behavioral and emotional characteristics:

Personality embraces a person's moods, attitudes, opinions, motivations, and style of thinking, perceiving, speaking, and acting. It is part of what makes each individual distinct. Theories of personality have existed in most cultures and throughout most of recorded history. The ancient Greeks used their ideas about physiology to account for differences and similarities in temperament. In the 18th century Immanuel Kant, Charles-Louis Montesquieu, and Giambattista Vico proposed ways of understanding individual and group differences; in the early 20th century Ernst Kretschmer and the psychoanalysts Sigmund Freud, Alfred Adler, and Carl Jung offered competing personality theories. Freud's model rested on the power of psychosexual drives as mediated by the structural components of the id, ego, and superego and the interplay of conscious and unconscious motives particularly important were the array of defense mechanisms an individual employed. Jung, like Freud, emphasized unconscious motives but de-emphasized sexuality and advanced a typal theory that classified people as introverts and extraverts; he further claimed that an individual personality was a persona drawn from the collective unconscious, a pool of inherited memories. Later theories by Erik H. Erikson, Gordon W. Allport, and Carl R. Rogers were also influential. Contemporary personality studies tend to be empirical and less theoretically sweeping and tend to emphasize personal identity and development. Personality traits are usually seen as the product of both genetic predisposition and experience.

Personality is made up the characteristic patterns of thoughts, feelings, and behaviours that make a person unique. Personality arises from within the individual and remains fairly consistent throughout life.

How to Improve Personality

We all believe our personalities are excellent. It is simply our behavior occasionally goes just a little off track, therefore when we speak about enhancing our personality, we have been really seeking at ways we all interact with the entire world and one another and our personal responsibility for the way we react for any situation and just how we exhibit our states and our feelings.

We may find that at some time in life, we have been struggling to work with stuff. We could be feeling anxious or emotional or angry or upset with other bad people at work place. The very first point to make is that we are in charge of outbursts, nobody else is. No-one can make you angry or make you behave in a certain way. So, we could decide the way we respond to the people close to us and we need to make much better decisions.

Components of Personality

While there are many different theories of personality, the first step is to understand exactly what is meant by the term personality. A brief definition would be that Personality is made up of the characteristic patterns of thoughts, feelings and behaviors that make a person unique. In addition to this, personality arises from within the individual and remains fairly consistent throughout life.

Big Five Personality Characteristics

The Big Five Personality Characteristics are a synthesis of the trait theories of personality developed by Cattell and Eysenck. Trait theories of personality attempt to describe a person with as few adjectives as possible. They are often used in a corporate setting or in job interviews to determine if a prospective employee will be able to work well in a company. Just what traits a person falls under are determined through analysis of a questionnaire that the person fills out.

The Big Five are commonly used because they combine the best of Cattell's comprehensive list of personality traits with the best of Eysenck's concise list, without resorting to redundancy. Listed with their corrolaries, they are:

- Extroversion—Introversion
- Neuroticism—Stability
- Agreeableness—Antagonism
- Conscientiousness—Undirectedness
- Openness—Nonopenness < br>

Psychologists claim to use the Big 5 for reasons more objective. They claim that factor analysis, a statistical method to determine correlation, detected these five trait clusters as being strongly internally correlated and not strongly correlated with one another. Each of the Big 5 is thus actually a set of smaller traits, called facets that are statistically linked.

The six facets of each of the big five are as follows:

- Extraversion:
  - Gregariousness
  - Activity Level
  - Assertiveness
  - Excitement Seeking
  - Positive Emotions
  - Warmth

- Agreeableness:
  - Straight forwardness
  - Trust
  - Altruism
  - Modesty
  - Tendermindedness
  - Compliance
- Conscientiousness:
  - Self-discipline
  - Dutifulness
  - Competence
  - Order
  - Deliberation
  - Achievement striving
- Neuroticism:
  - Anxiety
  - Self-consciousness
  - Depression
  - Vulnerability
  - Impulsiveness
  - Angry hostility
- Openness to new experience:
  - Fantasy
  - Aesthetics
  - Feelings
  - Ideas
  - Actions
  - Values

Personality Characteristics of People

Personality maybe defined as the visible aspect of one's character as he impresses it upon others. Thus, personality characteristics of people are what are visibly obvious in their behaviour towards others. Though the term 'personality' can have a legion of implications, however, let us only deal with the connotation. There have been various studies done to understand and classify different types of personalities and personality characteristics on the basis of behavior, ways of reacting in stressful situations and on the basis of several other guidelines, thus, often leading up to personality types.

So before you go running to take that personality test to find out who or what you are, let's stop to think of something. Nothing quite sums up the personality traits of a person than their facial expressions in different situations, as at the end of it all, facial expressions are a reflection of what a person is really thinking and feeling. To get an idea, just look around and you will see a myriad of people with different facial expressions all around.

Yet surprisingly, we have not done any experiment to categorize people on the basis of facial expression pointer. Hence, here is a quick review of how to recognize different personality types on the basis of his facial expression, and on the way he or she would behave in an office environment. And of course, this is the take on different types of personalities, a personal perspective on peculiar personalities. It will be cleared on reading the following characteristics:

Personality Characteristics

The Overbearing Type: Typical personality traits of the overbearing type include a deep look of concern that mars their eyes, which is often accompanied by knitted eyebrows and an ever ready nodding head while speaking to seniors. A slightly toned down version of narcissistic personality disorder, these people wish to impose themselves on others in every situation and are incessantly 'brimming with ideas'. Harnessing the values and traits inherited from their identical overbearing grandmothers, they wish to prove themselves right every step of the way. The kind that would make eager beavers seem like lethargic sloths, they open the door before opportunity can knock on it. Hence, although they may be a valuable asset in an office, often winning the affections of bosses and head honchos, they remain a species that colleagues and juniors very consciously try to avoid.

The Pacific Type: An ever pleasant smile and bright expressive eyes, accompanied by a 'ladies first' kind of rare-to-find-today courteous gentlemanly manners lands a person in this category. They are characterized by three 'C's - calm, composed and containing common sense. Although these personality traits make them better husbands than bosses, a relatively laid back attitude, accompanied by a non-dominating nature yet in possession of good decision-making skills, would categorize a person as being the pacific type - that is, someone who is calm on the surface but moving underneath, just like the Pacific ocean.

The Insouciant Type: If you find a person with a dreamy look in his eyes, which are half drooping with a 'devil-may-care or probably even he doesn't' attitude, then you've landed yourself with the insouciant type. These traits are typical of those seen at the other end of the spectrum of type A personality traits. They seem to be good listeners but beware, for they only seem to be listening, when they actually might be dreaming of the latest Lamborghini model. Looking like they are perennially deprived of sleep, they cast out periodic yawns which are dangerously contagious. Hence, needless to say, they aren't exactly an asset for any firm that is out to make profit. In other words, they are natures own counter-products to the overbearing type.

The Talkative Type: You may never get around to seeing their face because; you are way too preoccupied with trying to figure out a way of getting them to shut up. They brag, blabber, boast and back bite. Lawyers, secretaries, mothers-in-law all these are typically fall under this category. The energy that the brain is supposed to expend on coming up with ideas, is instead expended on overuse and abuse of the vocal cords. So, although not exactly an asset in the office they are indispensable for two reasons. Firstly, they are the ones that are often given the unenviable task of doing presentations in the office with colorful slides that hurt the eyes, and loud and long words that hurt the ears. Also, they form the vital group of people that linger around the water cooler all day and pass on the office gossip. So, as an afterthought, they are an evil necessity as offices without these personality types would be truly boring.

The Jittery Type: When it comes to the jittery type, nervousness is in their blood. Typical personality characteristics would include that even in freezing temperatures; beads of sweat are seen around their temple area. They have flickering eyes, quivering lips and an overall look of a person with the secret knowledge of a huge impending calamity. Their shirts or pants are often coffee stained as a result of the ceaseless tremors of their hands. They seem like the types who've been bullied to a pulp during their childhood, which has scarred them for life, thus exemplifying the need for personality and health correlation. They invite pitiful glances without even asking for any. Hence, although they may be exceptionally bright, their practically nonexistent PR skills will more often than not land them with a back office job.

So here was a sneak peek through a rare-view mirror at how to blue print peculiar personality characteristics that one comes across every day. So instead of falling back on back-breaking books on psychology, why not make your own classification to typecast different personalities because at the end of the day, it's all about what classification suits you best! So you can always come up with your own taxonomies, as long as you have the tact-xonomy, along with the patients and patience, to do so.

The big-five model of personality is an advance on preceding models insofar as it was founded on reappraisals of Cattell's sixteen factor model of personality, as measured by the 16 Personality Factor Questionnaire, or the "16PF".

It is true that there is a trend in personality psychology to regard the five-factor approach as an absolutely fundamental model, but its theoretical basis may leave a lot to be desired.

However, the five factors that have now generally accepted as encapsulating the five-factor model are those defined by Costa and McCrae as "Openness to Experience", "Conscientiousness", "Extraversion", "Agreeableness", and "Neuroticism"-easily remembered by the acronym "OCEAN".

An extensive criticism of the big-five approach has been put forward by Block which seriously questions the legitimacy of the approach of advocates of the big-five model. Whether or not the big-five model is theoretically sound can be determined by looking at how it came about - this is part of Block's critical approach.

The big-five model was derived from work by Cattell. Cattell was a proponent of the "lexical hypothesis" of personality theory, which states that terms describing all aspects of human personality that are important would have found their way into the language. On this rationale, Cattell reduced a list of 4,504 "non-judgemental trait-names" by Allport and Odbert and a series of his own terms used to describe conditions discovered by psychologists to 35 bipolar rating dimensions which he called the "standard reduced personality sphere". Cattell's model consists of 12 primary factors which emerged from his factor analyses. However, Block proposes that in making use of his own judgement and prior theoretical notions to decide upon his set of dimensions, Cattell may have entailed that his primary factors emerged from his analysis. This is an example of what is known as "prestructuring", which seems to be a running theme in Block's criticisms of the big-five model's development.

The big-five model really came about because of work by Tupes and Christal who factor analysed the results of Cattell's 16PF applied to 8 samples of young Air Force officers. In these analyses, 6 had 8 factors extracted; another had 5 factors, and the last, 12 factors. Tupes and Christal's claim was that "five fairly strong and recurrent factors emerged". Block, however, does not think that this recurrence is so striking because Tupes and Christal used the labourious centroid method of factor analysis for their first analysis, which extracted 8 factors, and this was rotated subjectively to make three of these factors "unimportant". Furthermore, their subsequent analyses used the less labourious multiple-group method, where the variables were grouped into five subsets prestructured so as to correspond to the five rotated factors decided upon in the first analysis. In other words, they were designed to conform to the structure of the first analysis and factors other than these five were residualised. More importantly, according to Block, was that Tupes and Christal may have prestructured their analysis in that it was based on the Cattellian variables, which contains many semantically related variables. Rating scales that contain many synonyms will almost always generate a mathematical factor. Block suggests that the factorial equivalence of many of the variables found in the Tupes and Christal study may only represent reliable and coherent scorings of the rater, not underlying factors. However, the reliability of the raters are also questioned on the basis that they received only very little training (about two hours at the most) and did not know their peers they were rating for very long. The final criticism of Tupes and Christal given by Block is that they named their five factors on the basis of a study by French without any elaboration on what they mean. They called them "Surgency or Extraversion", "Agreeableness", "Dependability", "Emotional Stability, or the Opposite of Emotionality", and "Culture", which Block understandably calls "allusive". Block also makes the point that French's study was based on many studies that were conceptually or methodologically insufficient.

The next advance in the big-five theory was made by Norman. On the basis of Tupes and Christal's factors, Norman selected 20 of their Cattellian variables - four for each factor. He then used these variables for undergraduate peer ratings, which were then factor, analysed. Block raises the very valid point that five factors obtained were a "forgone conclusion" due to a prestructured set of variables. However, Norman's study is seen as empirical support for the big-five structure. In addition to this study, Norman wanted to find any previously omitted or new personality terms, so he added 175 single-word descriptors to Allport and Odbert's list and used his own decision rules to exclude terms he judged quantitative, evaluative, ambiguous, metaphorical, vague, difficult, obscure, little-known, anatomical, physical, grooming characteristics, temporary states, or social rules. Norman only included those terms he deemed "stable traits", and after having these rated by undergraduates on their knowledge of these words; he was left with 1,431 terms. These terms were assigned to the positive or negative poles of his five dimensions and Norman further formulated what he judged to be semantic clusters within each pole.

Norman obtained 75 clusters, which were used by Goldberg in addition to some adjectives of his own to try to go beyond Cattell's work. Goldberg also found five factors in his studies, but this is hardly surprising given the use of Norman's clusters which were built around five factors in the first place - more prestructuring. It seems that the big-five models are based largely on the results of prestructured analyses, which does nothing to suggest that the model is a good advancement of the understanding of personality.

The big-five model has been derived largely from the lexical hypothesis, which Block does not think is totally sound. Block suggests that it is a "psychologically insufficient" hypothesis, drawing on the observation of McCrae and Costa that psychologists have uncovered important aspects of personality that were not encoded in the language. They make the good point that there is no reason to suspect that simple analysis of common English terms for parts of the body will provide an adequate basis for the science of anatomy, so why should this be true of personality, which is after all a much more intangible science? Block also adds that there are many aspects of personality that cannot be captured with a single-word term, such as "nasty to individuals of lower rank".

A further problem with the background of the big-five model is that Goldberg used undergraduates to determine if his stable trait descriptors were understandable by laypersons. Block suggests that undergraduates are not the best judge of this and it would be better to use a more expert judgement to avoid losing fairly unheard of, but possibly vitally important descriptors. He says that novices do not make the same discriminations that experts do.

Yet another problem for the idea of the big-five factors is that they have repeatedly been found to be non-orthogonal and correlate with each other. For instance, Goldberg's 100-item five-factor marker set is problematic. Although it produces five nearly orthogonal factors when data is taken from self-ratings and ratings of liked others, it does not produce orthogonal factors at all if the sample also includes ratings of disliked others. There are intercorrelations between Agreeableness, Conscientiousness, Emotional Stability, and Culture/Intellect/ Openness to Experience in the 30s to 50s. This just goes to show how the nature of the sample can affect the results of a factor analysis.

Finally, Costa and McCrae's questionnaire for testing their interpretation of the big-five factors has not escaped the criticism of Block. Costa and McCrae had initially found three clusters in Cattell's 16PF - Extraversion, Neuroticism, and also, a third and not well represented factor they termed "Openness to Experience". They then created a test using the 16PF variables and 6 additional ones for measuring Openness to Experience and found that their third factor was now better represented.

From this they constructed a new questionnaire based on their "NEO trait model". Block criticised the fact that this was not based on theory, just their personal thinking about how they should articulate their three domains.

However, after becoming influenced by the five-factor approach, Costa and McCrae wished to construct a questionnaire for the big-five which would integrate their NEO trait model. To do this they selected adjectives for their previously unconceptualised factors of Agreeableness and Conscientiousness from Goldberg. For the fifth factor, they tried to include aspects of Openness to Experience that they believed were underrepresented by Goldberg. Their factor analysis found five factors, the fifth of which now looked more like Openness to Experience than Goldberg's Intellect dimension. This seems to be a case of more prestructuring and indeed, even Costa and McCrae admit that their results may have been due to bias in the items that picked. Block rejects Costa and McCrae's claims that they showed that the five-factor model could encompass their NEO with Agreeableness and Conscientiousness being orthogonal to the NEO dimensions, instead suggesting that all they did was link two "nominally equivalent and predictably related lexical factors". Costa and McCrae's NEO Personality Inventory (NEO-PI) "grafted" the extra big-five factors onto the NEO Inventory and was found by Costa and McCrae to yield five-factors in analysis. However, Block suggests that this should not be seen as yet more validation of the big-five model, once again due to prestructuring. Even the latest version of Costa and McCrae's questionnaire, the Revised NEO Personality Inventory fails to validate the big-five model. In fact, it seems worse than it was before the revision as the Neuroticism and Conscientiousness scales have a -0.53 correlation, with a 0.40 correlation between the Extraversion and Openness to Experience scales. Furthermore, Parker found that the five factors did not emerge as expected. Block compounds the defeat of Costa and McCrae's approach as he sees that they allow their model, rather than eigenvalues or scree tests, to determine the number of factors they extract.

Advocates of the five-factor approach seem to be allowing "fiveness" to blind them somewhat to other possibilities. In addition to this, the big-five also fails to provide an adequate description of personality. Firstly, there is not enough clarity over what the factors actually mean. For instance, Costa and McCrae see "warmth" as a facet of Extraversion, while Goldberg sees it part of Agreeableness. Also, "Impulsivity" is classed as being part of Neuroticism by Costa and McCrae, and as being part of Extraversion by Goldberg. This shows that these factors are not clear, and perhaps this lack of clarity comes from the five factors being rather broad with fuzzy overlapping boundaries. This is something else that a good account of personality should not have. Summarising all the aspects of personality under five broad categories creates a great loss of information and no way of looking at more subtle traits. According to Block, a good model of personality also should not rely on factor analysis to decide the concepts of the model, and should also come from a conceptual language decided upon by experts rather than novices. The big-five model seems to fail on all these points and does not seem to have a solid theoretical background. The major problem seems to be that its apparent supporting evidence sufferers chronically from prestructuring. Therefore, the big-five does not make any advances in getting towards an understanding of what makes up personality.

Cattell's sixteen-factor model	HowCreativeAre You	projective personality test	personality test
Shave your head	Psychology is circular	Conscientiousness	Fibonacci like recurrence relations
introversion	extroversion	Gregariousness	Antagonism
characteristic	I hate you. Please don't leave me.	Personality disorders	stability

How To Meet Girls	I shaved my head last night	Hans Eysenck	Openness
job interview	Trait Theory	Big Five	

Contributions and Limitations of Cattell's Sixteen Personality Factor Model

Personality traits and scales used to measure traits are numerous and commonality amongst the traits and scales is often difficult to obtain. To curb the confusion, many personality psychologists have attempted to develop a common taxonomy. A notable attempt at developing a common taxonomy is Cattell's Sixteen Personality Factor Model based upon personality adjectives taken from the natural language. Although Cattell contributed much to the use of factor analysis in his pursuit of a common trait language his theory has not been successfully replicated.

Science has always strived to develop a methodology through which questions are answered using a common set of principles; psychology is no different. In an effort to understand differing personalities in humans, Raymond Bernard Cattell maintained the belief that a common taxonomy could be developed to explain such differences.

Cattell's scholarly training began at an early age when he was awarded admission to King's College at Cambridge University where he graduated with a Bachelor of Science in Chemistry in 1926. According to personal accounts, Cattell's socialist attitudes, paired with interests developed after attending a Cyril Burt lecture in the same year, turned his attention to the study of psychology, still regarded as a philosophy. Following the completion of his doctorate studies of psychology in 1929 Cattell lectured at the University at Exeter where, in 1930, he made his first contribution to the science of psychology with the Cattell Intelligence Tests. During fellowship studies in 1932, he turned his attention to the measurement of personality focusing of the understanding of economic, social and moral problems and how objective psychological research on moral decision could aid such problems. Cattell's most renowned contribution to the science of psychology also pertains to the study of personality. Cattell's 16 Personality Factor Model aims to construct a common taxonomy of traits using a lexical approach to narrow natural language to standard applicable personality adjectives. Though his theory has never been replicated, his contributions to factor analysis have been exceedingly valuable to the study of psychology.

Origins of the 16 Personality Factor Model

In developing a common taxonomy of traits for the 16 Personality Factor model, Cattell relied heavily on the previous work of scientists in the field. Previous development of a list of personality descriptors by Allport and Odbert in 1936, and Baumgarten's similar work in German in 1933, focused on a lexical approach to the dimensions of personality. Since psychology, like most other sciences, requires a descriptive model to be effective, the construction of a common taxonomy is necessary to successful in explaining personality simplistically. Already focused on the understanding of personality as it pertains to psychology, Cattell set out to narrow the work already completed by his predecessors. The goal of the research is to achieve integration as it relates to language and personality, that is, to identify the personality relevant adjectives in the language relating to specific traits.

The lexical approach to language is creates the foundation of a shared taxonomy of natural language of personality description. Historically, psychologists relied on such natural language to aid in the identification of personality attributes for such taxonomy. The first step in such a process was to narrow all adjectives within a language to those relating to personality descriptions, as it provided the researchers with a base guiding such a lexical approach. When working with a limited set of variables or adjectives within a language progressed from spoken word as it evolved throughout its progression. Since there are finite sets of adjectives in a language, the narrowing of the variables into base personality categories becomes necessary as multiple adjectives can express similar meanings within the language.

In the process of developing taxonomy, a process that had taken predecessors sixty years up to this point, Allport and Odbert systematized thousands of personality attributes in 1936. They recognized four categories of adjectives in developing the taxonomy including personality traits, temporary states highly evaluative judgments of personally conduct and reputation, and physical characteristics. Personality traits are defined as "generalized and personalized determining tendencies--consistent and stable modes of an individual's adjustment to their environment" as stated by Allport and Odbert in their research. Each adjective relative to personality falls within one of the previous categories to aid in the identification of major personality categories and creates a primitive taxonomy, which many psychologists and researchers would elaborate and build upon later. When Norman divided the same limited set of adjectives into seven categories where all mutually were exclusive. Despite this, work from both parties have been classified as containing ambiguous category boundaries, resulting in the general conviction that such boundaries should be abolished and the work has less significance than the earlier judgment.

Factor Analysis

Introduced and established by Pearson in 1901 and Spearman three years thereafter, factor analysis is a process by which large clusters and grouping of data are replaced and represented by factors in the equation. As variables are reduced to factors, relationships between the factors begin to define the relationships in the variables they represent. In the early stages of the process' development, there was little widespread use due largely in part to the immense amount of hand calculations required to determine accurate results, often spanning periods of several months. Later on a mathematical foundation would be developed aiding in the process and contributing to the later popularity of the methodology. In present day, the power of super computers makes the use of factor analysis a simplistic process compared to the 1900's when only the devoted researchers could use it to accurately attain results.

In performing a factor analysis, the single most import factor to consider is the selection of variables as considerations such as domain, where a single domain results in the highest accuracy, and other representative variables related to a single domain would provide a more accurate outcome. Exploratory factor analysis governs a single domain while confirmatory factor analysis, often less accurate and more difficult to calculate, governs several domains. In terms of variables, it is unlikely to see a factor analysis with fewer than 50 variables. In those situations, another statistical equation may be a better, easier consideration to process the information. A standard sample size for such a function would range between 500 to 1,000 participants.

Cattell, another champion of the factor analysis methodology, believed that there are three major sources of data when it comes to research concerning personality traits. L-Data, also referred to as the life record, could include actual records of a person's behavior in society such as court records. Cattell, however, gathered the majority of L-Data from ratings given by peers. Self -rating questionnaires, also known as Q-Data, gathered data by allowing participants to assess their own behaviors .The third source of Cattell's data the objective test, also known as T-Data, created a unique situation in which the subject is unaware of the personality trait being measured.

With the intent of generality, Cattell's sample population was representative of several age groups including adolescents, adults and children as well as representing several countries including the U.S., Britain, Australia, New Zealand, France, Italy, Germany, Mexico, Brazil, Argentina, India, and Japan.

Through factor analysis, Cattell identified what he referred to as surface and source traits. Surface traits represent clusters of correlated variables and source traits represent the underlying structure of the personality. Cattell considered source traits much more important in understanding personality than surface traits. The identified source traits became the primary basis for the 16 PF Model.

The 16 Personality Factor Model aims to measure personality based upon sixteen source traits. Table summarizes the surface traits as descriptors in relation to source traits within a high and low range.

Critical Review

Although Cattell contributed much to personality research through the use of factor analysis his theory is greatly criticized. The most apparent criticism of Cattell's 16 Personality Factor Model is the fact that despite many attempts his theory has never been entirely replicated. In 1971, Howarth and Brown's factor analysis of the 16 Personality Factor Model found 10 factors that failed to relate to items present in the model. Howarth and Brown concluded, "That the 16 PF does not measure the factors which it purports to measure at a primary level Studies conducted by Sell and by Eysenck and Eysenck also failed to verify the 16 Personality Factor Model's primary level. Also, the reliability of Cattell's self-report data has also been questioned by researchers.

Cattell and colleagues responded to the critics by maintaining the stance that the reason the studies were not successful at replicating the primary structure of the 16 Personality Factor model was because the studies were not conducted according to Cattell's methodology. However, using Cattell's exact methodology, Kline and Barrett, only were able to verify four of sixteen primary factors.

In response to Eysenck's criticism, Cattell, himself, published the results of his own factor analysis of the 16 Personality Factor Models, which also failed to verify the hypothesized primary factors.

Despite all the criticism of Cattell's hypothesis, his empirical findings lead the way for investigation and later discovery of the 'Big Five' dimensions of personality. Here, this is to mention that Fiske and Tupes and Christal had simplified Cattell's variables to five recurrent factors which are known as extraversion or surgency, agreeableness, consciousness, emotional stability and intellect.

Cattell's Sixteen Personality Factor Model has been greatly criticized by many researchers, mainly because of the inability of replication. More than likely, during Cattell's factor analysis errors in computation occurred resulting in skewed data, thus the inability to replicate. Since, computer programs for factor analysis did not exist during Cattell's time and calculations were done by hand it is not surprising that some errors occurred. However, through investigation into to the validity of Cattell's model researchers did discover the Big Five Factors, which have been monumental in understanding personality, as we know it today.

Some of the fundamental characteristics of personality include:

Consistency - There is generally a recognizable order and regularity to behaviors. Essentially, people act in the same ways or similar ways in a variety of situations.

Psychological and physiological - Personality is a psychological construct, but research suggests that it is also influenced by biological processes and needs.

It impacts behaviors and actions - Personality does not just influence how we move and respond in our environment; it also causes us to act in certain ways.

Multiple expressions - Personality is displayed in more than just behavior. It can also be seen in our thoughts, feelings, close relationships and other social interactions.

2

Theories of Personality

There are a number of different theories about how personality develops. Different schools of thought in psychology influence many of these theories. Some of these major perspectives on personality include:

Type theories are the early perspectives on personality. These theories suggested that there are a limited number of "personality types" which are related to biological influences.

Trait theories viewed personality as the result of internal characteristics that are genetically based.

Psychodynamic theories of personality are heavily influenced by the work of Sigmund Freud, and emphasize the influence of the unconscious on personality. Psychodynamic theories include Sigmund Freud's psychosexual stage theory and Erik Erikson's stages of psychosocial development.

Behavioral theories suggest that personality is a result of interaction between the individual and the environment. Behavioral theorists study observable and measurable behaviors, rejecting theories that take internal thoughts and feelings into account. Behavioral theorists include B. F. Skinner and John B. Watson.

Humanist theories emphasize the importance of free will and individual experience in the development of personality. Humanist theorists include Carl Rogers and Abraham Maslow.

A number of different theories have emerged to explain different aspects of personality. Some theories focus on explaining how personality develops while others are concerned with individual differences in personality. The following are just a few of the major theories of personality proposed by different psychologists:

- Trait Theories
- Gordon Allport's dispositional perspective
- Hans Eysenck's three-trait model
- Myers-Briggs Types
- "Big Five" Personality Dimensions
- Psychoanalytic Theories:
- Freud's Theory of Psychosexual Development

Freud's theory of psychosexual development is of the best known personality theories, but also one of the most controversial

Erikson's Theory of Psychosocial Development

According to Erik Erikson, each stage plays a major role in the development of personality and psychological skills. During each stage, the individual faces a developmental crisis that serves as a turning point in development.

Horney's Theory of Neurotic Needs

Theorist Karen Horney developed a list of neurotic needs that arise from overusing coping strategies to deal with basic anxiety. Learn more about these neurotic needs described by Horney.

Behavioral Theories:

Classical Conditioning: Classical conditioning is one of the best-known concepts of behavioral learning theory. In this type of conditioning, a neutral stimulus is paired with a naturally occurring response. Once an association has been formed, the previously neutral stimulus will come to evoke the response.

Operant Conditioning: Operant conditioning is one of the fundamental concepts in behavioral psychology. This process involves strengthening or weakening a behavior using reinforcement and punishment.

Humanist Theories:

Maslow's Hierarchy of Needs: Abraham Maslow's hierarchy of needs emphasizes the importance of self-actualization and is often pictured as a pyramid. The bases of the pyramid consist of basic survival needs, while the top of the pyramid is focused on self-actualizing needs.

Personality Disorders:

An estimated 10 to 15% of adults in the United States experience symptoms of at least one personality disorder. What are personality disorders? A personality disorder is a chronic and pervasive mental disorder that affects thoughts, behaviors, and interpersonal functioning. The DSM-IV currently lists 10 different personality disorders, which you can learn more about in the following article: Overview of Personality Disorders

Personality Tests:

You can find a number of personality tests here on the About Psychology site. These tests and quizzes are designed to give readers an idea of how formal assessments are used. However, these personality tests are not intended for use in assessment or diagnosis.

Improving social abilities: The concept covers a wider field than self-development or self-help: personal development also includes developing other people. This may take place through roles such as those of a teacher or mentor, either through a personal competency (such as the skill of certain managers in developing the potential of employees) or a professional service.

Beyond improving oneself and developing others, personal development is a field of practice and research. As a field of practice it includes personal development methods, learning programs, assessment systems, tools and techniques. As a field of research, personal development topics increasingly appear in scientific journals, higher education reviews, management journals and business books.

Any sort of development — whether economic, political, biological, organizational or personal — requires a framework if one wishes to know whether change has actually occurred. In the case of personal development, an individual often functions as the primary judge of improvement, but validation of objective improvement requires assessment using standard criteria. Personal development frameworks may include goals or benchmarks that define the end-points, strategies or plans for reaching goals, measurement and assessment of progress, levels or stages that define milestones along a development path, and a feedback system to provide information on changes.

Personal development in psychology: Psychology became linked to personal development, not with the psychoanalysis

Psychology is an academic and applied discipline that involves the scientific study of mental functions and behaviors. Psychology has the immediate goal of understanding individuals and groups by both establishing general principles and researching specific cases, and by many accounts it ultimately aims to benefit society. In this field, a professional practitioner or researcher is called a psychologist and can be classified as a social, behavioral, or cognitive scientist. Psychologists attempt to understand the role of mental functions in individual and social behavior, while also exploring the physiological and neurobiological processes that underlie certain cognitive functions and behaviors.

Psychologists explore concepts such as perception, cognition, attention, emotion, phenomenology, motivation, brain functioning, personality, behavior, and interpersonal relationships. Psychologists of diverse stripes also consider the unconscious mind. Psychologists employ empirical methods to infer causal and correlational relationships between psychosocial variables. In addition, or in opposition, to employing empirical and deductive methods, some—especially clinical and counseling psychologists—at times rely upon symbolic interpretation and other inductive techniques. Psychology has been described as a "hub science", with psychological findings linking to research and perspectives from the social sciences, natural sciences, medicine, and the humanities, such as philosophy.

While psychological knowledge is often applied to the assessment and treatment of mental health problems, it is also directed towards understanding and solving problems in many different spheres of human activity. The majority of psychologists are involved in some kind of therapeutic role, practicing in clinical, counseling, or school settings. Many do scientific research on a wide range of topics related to mental processes and behavior, and typically work in university psychology departments or teach in other academic settings (e.g., medical schools, hospitals). Some are employed in industrial and organizational settings, or in other areas such as human development and aging, sports, health, and the media, as well as in forensic investigation and other aspects of law.

Psychoanalysis is a psychological and psychotherapeutic theory founded in the late 19th century by Austrian neurologist Sigmund Freud. Since then, psychoanalysis has expanded, been criticized and developed in different directions, mostly by some of Freud's colleagues and students, such as Alfred Adler, Carl Gustav Jung and Wilhelm Reich, and later by neo-Freudians such as Erich Fromm, Karen Horney, Harry Stack Sullivan and Jacques Lacan.

The basic tenets of psychoanalysis include the following:

- Beside the inherited constitution of personality, a person's development is determined by events in early childhood;
- Human behavior, experience, and cognition are largely determined by irrational drives;
- Those drives are largely unconscious;
- Attempts to bring those drives into awareness meet psychological resistance in the form of defense mechanisms;

Conflicts between conscious and unconscious material can result in mental disturbances such as neurosis, neurotic traits, anxiety, depression etc.

The liberation from the effects of the unconscious material is achieved through bringing this material into the conscious mind.

Under the broad umbrella of psychoanalysis there are at least 22 theoretical orientations regarding human mental development. The various approaches in treatment called "psychoanalysis" vary as much as the theories do. The term also refers to a method of studying child development.

Freudian psychoanalysis refers to a specific type of treatment in which the "analysand" verbalizes thoughts, including free associations, fantasies, and dreams, from which the analyst induces the unconscious conflicts causing the patient's symptoms and character problems, and interprets them for the patient to create insight for resolution of the problems. The analyst confronts and clarifies the patient's pathological defenses, wishes and guilt. Through the analysis of conflicts, including those contributing to resistance and those involving transference onto the analyst of distorted reactions, psychoanalytic treatment can hypothesize how patients unconsciously are their own worst enemies: how unconscious, symbolic reactions that have been stimulated by experience are causing symptoms.

Psychoanalysis has been criticized on numerous fronts, including the view that it constitutes pseudoscience, but it remains influential within psychiatry.

Adler making the important points those aspirations looks forward and do not limit themselves to unconscious drives or to childhood experiences. He also originated the concepts of lifestyle and of self image, a concept that influenced management under the heading of work-life balance.

Carl Gustav Jung made contributions to personal development with his concept of individuation, which he saw as the drive of the individual to achieve the wholeness and balance of the Self.

Daniel Levinson included a sociological perspective. Levinson proposed that personal development come under the influence — throughout life — of aspirations, which he called "the Dream": Whatever the nature of his Dream, a young man has the developmental task of giving it greater definition and finding ways to live it out. It makes a great difference in his growth whether his initial life structure is consonant with and infused by the Dream. If the Dream remains unconnected to his life it may simply die, and with it his sense of aliveness and purpose.

Levinson's model of seven life-stages has been considerably modified due to sociological changes in the lifecycle

Research on success in reaching goals suggested that self-efficacy best explains why people with the same level of knowledge and skills get very different results. According to Bandera self-confidence functions as a powerful predictor of success because:

- It makes you expect to succeed;
- It allows you take risks and set challenging goals;
- It helps you keep trying if at first you don't succeed;
- It helps you control emotions and fears when the going gets rough.

Martin Seligman proposed a new focus: on healthy individuals rather than on pathology: - It has been discovered that there is a set of human strengths that are the most likely buffers against mental illness: courage, optimism, interpersonal skill, work ethic, hope, honesty and perseverance. Much of the task of prevention will be to create a science of human strength whose mission will be to foster these virtues in young people.

What is Personality Development, How to Improve Personality: Personality Development fundamentally means grooming and enhancing ones inner and outer personal to bring in the positive change for your life. Each and every individual features a distinct persona that could be polished, refined and developed. This technique includes improving communication, language speaking abilities and boosting ones confidence, developing certain hobbies, extending ones scope of understanding or skills, manners and studying fine etiquette, including grace and style for the way one appears, walks and talks and total imbibing oneself together with liveliness, positivity and peace.

A person's personality is actually an aggregate conglomeration of the decisions we make. You can find genetic, inherent natural and environmental factors which contribute for the development of our own personality.

If one looks up the definition of Personality development, you will find different explanations but the one that sums it up to an easy and effective understanding, is the following definition of Personality development which goes like this- 'an improvement in all spheres of an individual's life, be it with friends, in the office or in any other environment.'

This above listed definition is a clear and simple way for anyone to follow the meaning of personality development; in fact the definition has an underlying message about the importance of personality development for an individual as well. You wonder why and you wonder how? Well, by having the words 'improvement in all spheres of an individual's life', it lays importance to the fruitful aspects of having a personality-development process, this can be done via various means, be it a module to learn from, classes and seminars that one can attend, tips to follow and also, an on-going updating for personality development becomes necessary often enough, in the ever-changing world of today.

In any environment, one stands out if his or her personality is one that is bright and communicative. People skills are an integral part of any personality development programme. A pleasing personality is one that has a balance in behavior and is moderate in his or her reactions in public; while simultaneously maintain a friendly, approachable, yet sensible approach towards people. The developing personality is one, that has attributes that are enviable; the ability to perform well in public and be able to hold your own and make an impression on people with your skills socially and as a good personality, is something that cannot be matched up to by any other skill.

I say, developing personality, because personality development is easily possible with the right means and methods, but a personality is always learning and developing with experience and teachings. This is why the mention was made about updating on development of personality, because with constant updates, one keeps up with the trends that are changing like whirlwinds every now and then.

Yet, in a simple way of speech, personality development in general, as in, for a general outlook or a basis to perform with decent social skills, there is the scope for a concise and efficient personality development programme. This programme or module will allow for the personality to undergo development.

Personality Development quintessentially means enhancing and grooming one's outer and inner self to bring about a positive change to your life. Each individual has a distinct persona that can be developed, polished and refined.

This process includes boosting one's confidence, improving communication and language speaking abilities, widening ones scope of knowledge, developing certain hobbies or skills, learning fine etiquettes and manners, adding style and grace to the way one looks, talks and walks and overall imbibing oneself with positivity, liveliness and peace.

The whole process of this development takes place over a period of time. Even though there are many crash courses in personality development that are made available to people of all age groups, implementing this to your routine and bringing about a positive change in oneself takes a considerable amount of time. It is not necessary to join a personality development course; one can take a few tips and develop his or her own aura or charm.

- You may have heard this a million times "Think Positive". It works.
- Smile. And smile some more. It adds to your face value and to your personality as well.
- Read a few articles in the newspaper loudly. This will help in communicating fluently.
- Follow table manners and dining etiquettes
- Take good care of your health, dress well, be neat and organized
- Prepare a chart that mentions your strengths and weaknesses. Now concentrate on the latter and find ways to improve upon the same. Do not forget to strengthen your strengths.
- Spend some time alone concentrating on you and yourself alone.
- Practice meditation and yoga. It will help you develop inner peace and harmony that will reflect outside.
- Do not live a monotonous life. Be creative and do something new all the time. Nothing bigger than the joy of creative satisfaction.

Personality development is gaining more and more importance because it enables people to create a good impression about themselves on others; it helps them to build and develop relationships, helps in your career growth and also helps to improve your financial needs.

After all, personality development is nothing but a tool that helps you realize your capabilities and your strengths making you a stronger, a happier and a cheerful person.

Definition of Personality Development

Personality development is actually the development from the organized pattern of attitudes and behaviors which makes an individual distinctive. Personality development takes place by the continuous interaction of the temperament, environment and character. This is the Development of habitual designs of behaviour in adolescence and childhood.

You can find many theories of personality, however the very first step is always to understand precisely what is the term personality means.

A quick definition could be, personality is composed of the characteristic designs of feelings, behaviors and thoughts which make a person special. In addition personality occurs from within the person and stays fairly consistent during the whole life.

Tips for Personality Development: Personality is actually a number of characteristic feelings, behaviors and thoughts which are related to an individual. Personality development will be improvement in every sphere of a person's life, whether it is with friends, office or in some other environment. The vedantic idea of personality improvement is according to the concept of excellence of each and every soul and self-confidence to manifestation and realization of the inner knowledge. 5 dimensions take part in forming a famous personality. They are:

- Blissful self
- Energy self
- Mental self
- Physical self
- Intellectual self.

Importance of Personality Development: Let's not go with the bookish definition of personality development. Let me give you an example: You must have been to parties, get-togethers etc. Have you noticed that there are some people who are surrounded by lot of friends all the time and there are others who sit in the corner of the room having a drink alone? The difference between these two groups is mainly because of their different characters, thoughts and different personalities.

Personality is a blend of your temperament, emotion, principles and behavior. You start developing a personality when you are a child. This personality is later molded, based on the impact of various positive and negative factors in life. However, the significance of this understanding is that, you can always keep improving your personality. In other words, if you feel that there are some features in your personality that needs to be worked upon, then it can be done.

Personality development plays a significant role in three most important aspects of our life:

1) Social Relationships
2) Family Relationships
3) Professional Relationships

Social Relationships: Human being is a social animal. Acceptance from friends and society are very important for an individual, unless you are strong enough, not to care about others. However, at some point or the other, you do feel aloof and ignored. An individual with attractive personality has charisma, with magnetic power, to draw people towards him and make them know what he wants. Who doesn't feel great when you are recognized and accepted by the society? If you are shy and timid that shows lack of confidence in your personality.

Family Relationship: The home environment is warm and pleasant when a family member has vibrant and positive personality, full of fun and happiness. It's the pleasure and laughter that makes the home atmosphere to live on. The way you interact with your partner or parents or children, speak a lot about your character. A weak relationship shows lack of confidence and trust. If your family relationship is not strong, that shows in your other relationships as well.

Professional Relationship: Meeting deadlines, maintaining good relations with your boss, punctuality are some of the challenges that an individual has to face in the office. It's not the number of challenges that matters. What matters is how you cope and balance these challenges to benefit your job. If you get frustrated on simple matters, you would easily give up. Personality Development plays such a significant role in organizations today, that you will find Personality Development Training Courses for all employees, to develop their skills and raise their motivation level. With right personality and skills, it becomes easier to interact with clients, colleagues and bosses. And also helps to face the challenge with positive approach.

Approach for Personality Development: First step towards that perfect personality is to be honest to you. You need to analyze yourself and evaluate your existing personality. This will help you understand your shortcomings. Some important features of personality development are as follows:

- Cleanliness
- Honesty
- Unselfishness
- Courteous
- Respect
- Good Health
- Kindness

- Confidence
- Open Mindedness

It is not easy to train yourself and achieve above features in a day or two. You need to work hard. Take each quality one at a time and try to practice it in your daily life.

Any personality is first judged by its physical appearance, so cleanliness, your dressing style, the way you walk, how you groom etc is very important. 80% of your physical appearance can be improved just by a good posture. If you walk straight with your head held high, you automatically develop grace in your movement.

Nobody in this world is perfect. But if you wish to grow in your life, you need to make yourself shine by polishing your character and behavior. Remember, only practice can make a man perfect and help in transforming a personality into the one that can succeed in every aspect of a life.

Components of Personality Development: There are immense possibilities within you to develop your personality with strong character. You must recognize and understand them. The key to success is hidden within you. The wealth you have earned is not important but what is important is the means you have adopted for acquiring that wealth. Your personality, your capabilities, your thoughts and your ideals are all very important in determining your character. We will discuss some essential components of personality Development.

Be polite and speak sweetly: Sweetly spoken words have their own advantage and it is a major component in developing our personality. Society gives respect to a person who has control over his speech and everybody wants to conserve with a person who is sweet spoken. That is why we should always try and inculcate sweet and soft spoken words in our speech.

All actions are controlled by tongue or speech. They originate from speech. So it is essential that speech is controlled by mind. A person who is not honest by speech is considered dishonest in all respects. The man who has not control over his speech cannot be sure if he can exercise any control over his self while working. If a wrong thing is said in good style it also becomes effective. Then if you say a good thing in good way it has no comparison.

Keep away from lies, fraud and deception: G.D. Broadman, a great western thinker said – sow a work and get a habit, sow a habit and get a character, sow a character and get a fortune.

It is evident that the fate of a person is formed by the work he dies, the branch of a tree bends in the direction it is diverted. If a person once entangles him in lies and cunningness, then in spite of making countless efforts he can never get out of it and day by day he will keep moving toward downfall. We never improve our personality keeping these elements within us or a part our habits.

Once a man resorts to lies for self gains and self protection then he gradually becomes an expert in telling lies. Lies, treachery, manipulation and fraud are such aspects of our personality which will lead to our failure, insult and defamation at every step of life. It is necessary that we get rid of these negative shortcomings with firm determination to get success in life. It is an admitted fact that bad ideas develop more rapidly as compared to good ideas. Therefore, we must try and keep ourselves away from all kinds of bad ideas and their impact.

Believe in healthy competition: The path of all success is filled with thorns. There is no hindrance in the path of downfall, but no sooner we lift the first step towards progress we start facing difficulties. Only those people who have immense courage and enthusiasm can advance on this path. Those who have moved forward have become immortal. These who cross all the barriers, all the pains, sufferings and sorrows are also able to taste the sweet fruits of success.

To achieve success it is necessary that all the energies of body, mind and wealth are utilized in one direction only. When a nab m avoiding all temptations in life works towards one definite foal, he is bound to get total victory. When all energies of mind and soul are concentrated for a particular work then all sources become available for completion of that particular work.

The importance of positive attitude: Positive attitude is of great importance in our personality. Many people, in spite of having sufficient talent and qualification, do not progress and succeeds in life because of their narrow outlook. Nobody likes people who have narrow outlook. In life only that person is considered will behave who have positive and health bend of mind.

The real beauty of a man lies not in his physical appearance but in his work and good qualities. The ability and achievements of a person are judged by his mentality. That is why it is important that we improve our mental outlook. By adopting positive outlook we can get rid of mental dissatisfaction. This change will help you to contribute a lot in making your relations better and sweet with people as well as your personality will bright all over as like the stars in the sky.

Anger and irritation are your enemies: To be angry is part of human nature. Probably there is no man in this world who has never been angry for some or the other reason. But some people are such who become temperamentally angry. People who get angry and irritated habitually do not realize that this habit of theirs will be a hindrance in path of their personality development and they will become their own enemies.

Anger results in foolishness and insanity. There is a saying that hot temper is the main obstacle in completion of delicate genes of the body. Sometimes delicate tissues and veins even get burnt, outwardly we do not come to know that damage has been caused but we destroy many hidden treasures in our body in fire of anger.

Spare time for social service: If we want to be recognized in the society and want to reach the pinnacle of success, we should not keep out self aloof from society. Our deep feelings and responsibility towards society makes us good and responsible citizen of the society we live in. The more we work for society and the more we remain connected with it, the more famous we will become the fame will enhance our personality and enthusiasm. Always remember that society is ours. We need it. Live not only for yourself but for the society.

Develop your creative power: Our creative streak and positive propensity can pay us dividends. One who is engaged in constructive activity of one or the other kind is always happy. All men should channelize their energies in creative work. Be it agriculture, physical labour, daily wager, craftsmanship, office work, domestic chores or social work. Breach in creativity is a breach in peace and happiness.

Know the value of time: Every moment of our life is a golden moment. The door of progress and development remain open for the person who knows to utilize and make best use of these moments. He not only pushes away all the failures and gets victory over all setbacks and obstacle but also becomes very popular in society dye to his punctual behavior. On the other hand worthless pursuits or pass these moments carelessly, face depression and disappointment on all sphere victims of vicious circle of failures.

3

What is a Personality Disorder?

What is a personality disorder and what are the criteria for diagnosing a personality disorder?

According to the Diagnostic and Statistical Manual of Mental Disorders a personality disorder is "an enduring pattern of inner experience and behavior" that meets the following criteria:

Deviates from the expectations of the individual's culture is pervasive and inflexible has an onset in adolescence or early adulthood is stable over time leads to impairment or distress

Borderline personality disorder is one of 10 personality disorders that are recognized in the DSM-IV. Other personality disorders include: paranoid, schizoid, schizotypal, antisocial, histrionic, narcissistic, avoidant, dependent and obsessive-compulsive personality disorder.

In addition, the DSM-IV designates the category of Personality Disorder Not Otherwise Specified, or PD-NOS, for individuals who meet the general criteria for a personality disorder and have symptoms of one or more personality disorders but not enough to meet criteria for any specific personality disorder. A person may also be diagnosed with PD-NOS if they meet the general criteria for a personality disorder but have symptoms of a personality disorder that are not formally recognized in the DSM-IV at this time.

Personality disorders are a group of mental health conditions in which a person has a long-term pattern of behaviors, emotions, and thoughts that is very different from his or her culture's expectations. These behaviors interfere with the person's ability to function in relationships, work, or other settings.

According to the DSM-IV-TR, psychology's premier diagnostic reference, those with psychological personality disorders have traits that cause them to feel and behave in socially distressing ways. Depending on the specific disorder, these personalities are generally described in negative terms such as hostile, detached, needy, antisocial or obsessive. People with personality disorders typically experience discord and instability in many aspects of their lives, and most are prone to blame others for their problems.

While many other psychological disorders fluctuate in terms of symptom presence and intensity, as with normal personality, personality disorders typically remain relatively constant throughout life, although they do vary in severity from individual to individual.

Causes of Personality Disorders

Causes, incidence, and risk factors

Causes of personality disorders are unknown. Genetic and environmental factors are thought to play a role.

Mental health professionals categorize these disorders into the following types:

Symptoms

Symptoms vary widely depending on the type of personality disorder.

In general, personality disorders involve feelings, thoughts, and behaviors that do not adapt to a wide range of settings.

These patterns usually begin in adolescence and may lead to problems in social and work situations.

The severity of these conditions ranges from mild to severe.

Signs and tests

Personality disorders are diagnosed based on a psychological evaluation that assesses the history and severity of the symptoms.

Treatment

At first, people with these disorders usually do not seek treatment on their own. They tend to seek help once their behavior has caused severe problems in their relationships or work. They may also seek help when they are struggling with another mental health problem, such as a mood or substance abuse disorder.

Although personality disorders take time to treat, certain forms of talk therapy are helpful. In some cases, medications are a useful addition.

Expectations (prognosis)

Outlook varies. Some personality disorders improve greatly during middle age without any treatment. Others only improve slowly even with treatment.

Complications

Problems with relationships

Problems with school or work

Other mental health disorders

The combined and consistent patterns of emotion, thought and behavior that make an individual unique comprise personality. This aspect of self differs from mood in that mood is a more changeable, often situational response, whereas personality is a relatively stable, life-long combination of traits.

Think of the overall list of general characteristics that those who know you would use if asked to describe what you're like. Whether you are cheerful and optimistic or reclusive and brooding, these characteristics are a product of your heredity and early life experience, and are generally fixed by the time you reach adulthood.

There is still much to be learned about the causes of disorders in this class. Although the specific origins of personality disorders are yet unclear, studies have consistently shown child abuse and neglect as being antecedent risks to the development of personality disorders in adulthood. However, history of abuse is not evident for all patients diagnosed with a personality disorder.

Classification of Personality Disorders

There are currently 10 conditions that are considered personality disorders, some of which have very little in common. Mental health professionals typically group those personality disorder types that share characteristics into one of three clusters:

Cluster A Personality Disorders are those considered to be marked by odd, eccentric behavior. Paranoid (PPD), Schizoid and Schizotypal Personality Disorders are in this category.

Cluster B Personality Disorders are evidenced by dramatic, erratic behaviors and include Histrionic (HPD), Narcissistic (NPD), Antisocial (APD or ASPD) and Borderline (BPD) Personality Disorders.

Cluster C Personality Disorders are distinguished by the anxious, fearful behavior commonly seen in Obsessive-Compulsive (OCPD), Avoidant (APD) and Dependent (DPD) Personality Disorders.

Diagnosis of Personality Disorders

To be diagnosed with a disorder in this category, a psychologist or psychiatrist must evaluate the patient for the following:

Symptoms have been present for an extended period of time, are inflexible and pervasive, and are not a result of alcohol or drugs or another psychiatric disorder. The symptoms began to manifest in adolescence or early adulthood.

Symptoms cause significant distress or negative consequences in the person's life.

Symptoms are seen in at least two of the following areas:

- Thoughts
- Emotions
- Interpersonal functioning
- Impulse control

Prognosis for those with Personality Disorders

Although there's no cure for these conditions, therapy and medication can help many individuals manage their symptoms. However, those with personality disorders often have difficulty maintaining consistent psychiatric care due to poor relationships with medical and mental health professionals. Those suffering may refuse to take responsibility for their behavior or, as a manifestation of their illness, feel overly distrustful, deserving, or needy, taxing the doctor-patient relationship. These patients are also unlikely to faithfully follow a prescribed treatment regimen of regular psychiatric appointments and are often noncompliant with drug therapy. It may be small consolation, but the symptoms of some personality disorders do improve with age.

Antisocial personality disorder

Antisocial personality disorder is a mental health condition in which a person has a long-term pattern of manipulating, exploiting, or violating the rights of others. This behavior is often criminal.

Causes, incidence, and risk factors

Cause of antisocial personality disorder is unknown. Genetic factors and environmental factors, such as child abuse, are believed to contribute to the development of this condition. People with an antisocial or alcoholic parent are at increased risk. Far more men than women are affected. The condition is common among people who are in prison.

Fire-setting and cruelty to animals during childhood are linked to the development of antisocial personality.

Some doctors believe that psychopathic personality is the same disorder. Others believe that psychopathic personality is a similar but more severe disorder.

Symptoms

A person with antisocial personality disorder may:

- Be able to act witty and charming
- Be good at flattery and manipulating other people's emotions
- Break the law repeatedly
- Disregard the safety of self and others
- Have problems with substance abuse
- Lie, steal, and fight often
- Not show guilt or remorse
- Often be angry or arrogant

Signs and tests

Antisocial personality disorder is diagnosed based on a psychological evaluation that assesses the history and severity of symptoms. To be diagnosed with antisocial personality disorder, a person must have had conduct disorder during childhood.

Treatment

Antisocial personality disorder is one of the most difficult personality disorders to treat. People with this condition rarely seek treatment on their own. They may only start therapy when required to by a court.

Behavioral treatments, such as those that reward appropriate behavior and have negative consequences for illegal behavior, may hold the most promise. Certain forms of talk therapy are also being explored.

Persons with antisocial personality who have other disorders, such as a mood or substance disorder, are often treated for those problems as well.

Expectations (prognosis)

Symptoms tend to peak during the late teenage years and early 20s. They sometimes improve on their own by a person's 40s.

Complications

Complications can include imprisonment, drug abuse, violence, and suicide.

Avoidant personality disorder

Personality disorder - avoidant

Avoidant personality disorder is a mental health condition in which a person has a lifelong pattern of feeling very shy, inadequate, and sensitive to rejection.

Causes, incidence, and risk factors

Cause of avoidant personality disorder is unknown. Genes or a physical illness that changed the person's appearance may play a role. About 1% of the population has avoidant personality disorder.

Symptoms

People with this disorder cannot stop thinking about their own shortcomings. They form relationships with other people only if they believe they will not be rejected. Loss and rejection are so painful that these people choose to be lonely rather than risk trying to connect with others.

A person with avoidant personality disorder may:

- Be easily hurt when people criticize or disapprove of them
- Hold back too much inintimate relationships
- Be reluctant to become involved with people
- Avoid activities or jobs that involve contact with others
- Be shy in social situations out of fear of doing something wrong
- Make potential difficulties seem worse than they are
- Hold the view they are not good socially, not as good as other people, or unappealing

Signs and tests

Avoidant personality disorder is diagnosed based on a psychological evaluation that assesses the history and severity of the symptoms.

Treatment

Talk therapy is considered to be the most effective treatment for this condition. It helps persons with this disorder be less sensitive to rejection. Antidepressant drugs may be used in addition.

Expectations (prognosis)

People with this disorder may develop some ability to relate to others. With treatment this can be improved.

Complications

Without treatment, a person with avoidant personality disorder may lead a life of near or total isolation. They may go on to develop a second mental health disorder such as substance abuse or a mood disorder such as depression.

Borderline personality disorder

Borderline personality disorder (BPD) is a mental health condition in which a person has long-term patterns of unstable or turbulent emotions. These inner experiences often result in impulsive actions and chaotic relationships with other people.

Causes, incidence, and risk factors

Cause of borderline personality disorder is unknown. Genetic, family, and social factors are thought to play roles.

Risk factors for BPD include:

- Abandonment in childhood or adolescence
- Disrupted family life
- Poor communication in the family
- Sexual, physical, or emotional abuse

This personality disorder tends to occur more often in women and among hospitalized psychiatric patients.

Symptoms

Persons with BPD are often uncertain about their identity. As a result, their interests and values can change rapidly. They also tend to view things in terms of extremes, such as either all good or all bad. Their views of other people can change quickly. A person who is looked up to one day may be looked down on the next day. These suddenly shifting feelings often lead to intense and unstable relationships.

Other symptoms of BPD include:

- Intense fear of being abandoned
- Cannot tolerate being alone
- Frequent feelings of emptiness and boredom
- Frequent displays of inappropriate anger
- Impulsiveness, such as with substance abuse or sexual relationships
- Repeated crises and acts of self-injury, such as wrist cutting or overdosing

Signs and tests

BPD is diagnosed based on a psychological evaluation that assesses the history and severity of the symptoms.

Treatment

Individual talk therapy may successfully treat BPD. In addition, group therapy can sometimes be helpful.

Medications have less of a role in the treatment of BPD. But in some cases, they can improve mood swings and treat depression or other disorders that may occur with this condition.

Expectations (prognosis)

Outlook of treatment depends on how severe the condition is and whether the person is willing to accept help. With long-term talk therapy, the person often gradually improves.

Complications

- Depression
- Drug abuse
- Problems with work, family, and social relationships
- Suicide attempts and actual suicide

Dependent personality disorder

Dependent personality disorder is a mental health condition in which people depend too much on others to meet their emotional and physical needs.

Causes, incidence, and risk factors

Cause of dependent personality disorder is unknown. The disorder usually begins in childhood. It is one of the most common personality disorders and is equally common in men and women.

Symptoms

People with this disorder do not trust their own ability to make decisions. They may be very upset by separation and loss. They may go to great lengths, even suffering abuse, to stay in a relationship.

Symptoms of dependent personality disorder may include:

- Avoiding being alone
- Avoiding personal responsibility
- Becoming easily hurt by criticism or disapproval
- Becoming overly focused on fears of being abandoned
- Becoming very passive in relationships
- Feeling very upset or helpless when relationships end
- Having difficulty making decisions without support from others
- Having problems expressing disagreements with others

Signs and tests

Dependent personality disorder is diagnosed based on a psychological evaluation that assesses the history and severity of the symptoms.

Treatment

Talk therapy is considered to be the most effective treatment. The aim is to help people with this condition make more independent choices in life. Medicines may help treat other mental health conditions, such as anxiety or depression that occur along with this disorder.

Expectations (prognosis)

Improvements are usually seen only with long-term therapy.

Complications

- Alcohol or drug abuse
- Depression
- Increased likelihood of physical, emotional, or sexual abuse

Histrionic personality disorder

Histrionic personality disorder is a mental health condition in which people act in a very emotional and dramatic way that draws attention to them.

Causes, incidence, and risk factors

Cause of histrionic personality disorder is unknown. Genes and early childhood events may be responsible. It is diagnosed more often in women than in men. Doctors believe that more men may have the disorder than are diagnosed.

Histrionic personality disorder usually begins by late teens or early 20s.

Symptoms

People with this disorder are usually able to function at a high level and can be successful socially and at work.

Symptoms include:

- Acting or looking overly seductive
- Being easily influenced by other people
- Being overly concerned with their looks
- Being overly dramatic and emotional
- Being overly sensitive to criticism or disapproval
- Believing that relationships are more intimate than they actually are
- Blaming failure or disappointment on others
- Constantly seeking reassurance or approval
- Having a low tolerance for frustration or delayed gratification
- Needing to be the center of attention
- Quickly changing emotions, which may seem shallow to others

Signs and tests

Histrionic personality disorder is diagnosed based on a psychological evaluation that assesses the history and severity of the symptoms.

The health care provider can diagnose histrionic personality disorder by looking at the person's:

- Behavior
- History
- Overall appearance
- Psychological evaluation

Treatment

People with this condition often seek treatment when they experience depression or anxiety from failed romantic relationships or other conflicts with people. Medicine may help the symptoms. Talk therapy is the best treatment for the condition itself.

Expectations (prognosis)

Histrionic personality disorder can improve with talk therapy and sometimes medicines. Left untreated, it can cause problems in people's personal lives and prevent them doing their best at work.

Complications

Histrionic personality disorder may affect a person's social or romantic relationships. The person may be unable to cope with losses or failures. The person may change jobs often because of boredom and not being able to deal with frustration. A person with this disorder craves new things and excitement, which leads to risky situations. All of these factors may lead to a higher chance of depression.

Narcissistic personality disorder

Narcissistic personality disorder is a condition in which people have an excessive sense of self-importance, an extreme preoccupation with themselves, and lack of empathy for others.

Causes, incidence, and risk factors

Cause of this disorder is unknown. Early life experiences, such as particularly insensitive parenting, are thought to play a role in the development of this disorder.

Symptoms

A person with narcissistic personality disorder may:

- React to criticism with rage, shame, or humiliation
- Take advantage of other people to achieve his or her own goals
- Have excessive feelings of self-importance
- Exaggerate achievements and talents

- Be preoccupied with fantasies of success, power, beauty, intelligence, or ideal love
- Have unreasonable expectations of favorable treatment
- Need constant attention and admiration
- Disregard the feelings of others, and have little ability to feel empathy
- Have obsessive self-interest
- Pursue mainly selfish goals

Signs and tests

Narcissistic personality disorder is diagnosed based on a psychological evaluation that assesses the history and severity of the symptoms.

Treatment

Talk therapy may help the affected person relate to other people in a more positive and compassionate way.

Expectations (prognosis)

Outcome of treatment depends on the severity of the disorder.

Complications

- Alcohol or other drug dependence
- Relationship, work, and family problems

Obsessive-compulsive personality disorder

Obsessive-compulsive personality disorder (OCPD) is a mental health condition in which a person is preoccupied with rules, orderliness, and control.

Causes, incidence, and risk factors

OCPD tends to occur in families, so genes may be involved. A person's childhood and environment may also play roles.

This disorder can affect both men and women. It most often occurs in men.

Symptoms

OCPD has some of the same symptoms as obsessive-compulsive disorder (OCD). But people with OCD have unwanted thoughts, while people with OCPD believe that their thoughts are correct. In addition, OCD often begins in childhood while OCPD usually starts in the teen years or early 20s.

People with either OCPD or OCD are high achievers and feel a sense of urgency about their actions. They may become very upset if other people interfere with their rigid routines. They may not be able to express their anger directly. People with OCPD have feelings that they consider more appropriate, like anxiety or frustration.

A person with OCPD has symptoms of perfectionism that usually begin by early adulthood. This perfectionism may interfere with the person's ability to complete tasks, because their standards are so rigid. They may withdraw emotionally when they are not able to control a situation. This can interfere with their ability to solve problems and form close relationships.

Other signs of OCPD include:

- Over-devotion to work
- Not being able to throw things away, even when the objects have no value
- Lack of flexibility
- Lack of generosity
- Not wanting to allow other people to do things
- Not willing to show affection
- Preoccupation with details, rules, and lists

Signs and tests

OCPD is diagnosed based on a psychological evaluation

that assesses the history and severity of the symptoms.

Treatment

Medicines may help reduce anxiety and depression from OCPD. Talk therapy is thought to be the most effective treatment for OCPD. In some cases, medicines combined with talk therapy are more effective than either treatment alone.

Expectations (prognosis)

Outlook for OCPD tends to be better than that for other personality disorders. The rigidness and control of OCPD may prevent many of the complications such as drug abuse, which are common in other personality disorders.

The social isolation and difficulty handling anger that are common with OCPD may lead to depression and anxiety later in life.

Complications

- Anxiety
- Depression
- Difficulty moving forward in career situations
- Relationship difficulties

Paranoid personality disorder

Paranoid personality disorder is a mental health condition in which a person has a long-term pattern of distrust and suspicion of others, but does not have a full-blown psychotic disorder such as schizophrenia.

Causes, incidence, and risk factors

Cause of paranoid personality disorder is unknown. The disorder appears to be more common in families with psychotic disorders such as schizophrenia and delusional disorder. This suggests genes may be involved. Environmental factors may play a role as well.

The condition appears to be more common in men.

Symptoms

Persons with paranoid personality disorder are very suspicious of other people. As a result, they severely limit their social lives. They often feel that they are in danger and look for evidence to support their suspicions. They have trouble seeing that their distrustfulness is out of proportion to their environment.

Common symptoms include:

- Concern that other people have hidden motives
- Expectation that they will be exploited (used) by others
- Inability to work together with others
- Social isolation
- Detachment
- Hostility

Signs and tests

Paranoid personality disorder is diagnosed based on a psychological evaluation that assesses the history and severity of the symptoms.

Treatment

Treatment is difficult because people with this condition are often very suspicious of doctors. If treatment is accepted, talk therapy and medications can often be effective.

Expectations (prognosis)

Outlook usually depends on whether the person is willing

to accept help. Talk therapy and medications can sometimes reduce paranoia and limit its impact on the person's daily functioning.

Complications

- Extreme social isolation
- Problems with school or work

Schizoid personality disorder

Schizoid personality disorder is a mental health condition in which a person has a lifelong pattern of indifference to others and social isolation.

Causes, incidence, and risk factors

Cause of schizoid personality disorder is unknown. This disorder may be related to schizophrenia and shares many of the same risk factors.

Schizoid personality disorder is generally not as disabling as schizophrenia. It does not cause the disconnection from reality that occurs in untreated schizophrenia.

Symptoms

A person with schizoid personality disorder often:

- Appears aloof and detached
- Avoids social activities that involve emotional intimacy with other people
- Does not want or enjoy close relationships, even with family members

Signs and tests

Schizoid personality disorder is diagnosed based on a psychological evaluation that assesses the history and severity of the symptoms.

Treatment

People with this disorder rarely seek treatment, thus little is known about which treatments work. Talk therapy may not be effective because persons with schizoid personality disorder may have great difficulty forming an effective working relationship with a therapist.

One approach that appears to help is to put fewer demands for emotional closeness or intimacy on the person with this condition.

People with schizoid personality disorder often do well in relationships that do not focus on emotional closeness. They tend to be better at handling relationships that focus on work or intellectual activities and expectations.

Expectations (prognosis)

Schizoid personality disorder is a long-term (chronic) illness that usually does not improve much over time. Social isolation often prevents the person from seeking the help or support that might improve the condition.

Limiting expectations of emotional intimacy may help people with this condition make and keep connections with other people.

Schizotypal personality disorder

Schizotypal personality disorder is a mental health condition in which a person has trouble with relationships and disturbances in thought patterns, appearance, and behavior

Causes, incidence, and risk factors

Cause of schizotypal personality disorder is unknown. Genes are thought to be involved because this condition is more common in relatives of schizophrenics.

Symptoms

Schizotypal personality disorder should not be confused with schizophrenia. People with schizotypal personality disorder can have odd beliefs and behaviors. But unlike people with schizophrenia, they are not disconnected from reality and usually do not hallucinate. They also do not have delusions.

People with schizotypal personality disorder may be very disturbed. For example, they may also have unusual preoccupations and fears, such as fear of being monitored by government agencies.

More commonly, people with this disorder behave oddly and have unusual beliefs. They cling to these beliefs so strongly that they have difficulty forming and keeping close relationships.

People with this disorder may also have depression. A second personality disorder, such as paranoid personality disorder, is also common.

Common signs of schizotypal personality disorder include:

- Discomfort in social situations
- Inappropriate displays of feelings
- No close friends
- Odd behavior or appearance
- Odd beliefs, fantasies, or preoccupations
- Odd speech

Signs and tests

Schizotypal personality disorder is diagnosed based on a psychological evaluation that assesses the history and severity of the symptoms.

Treatment

Talk therapy is an important part of treatment. Social

skills training can help some people cope with social situations. Medicines may also be a helpful addition.

Expectations (prognosis)

Schizotypal personality disorder is usually a long-term (chronic) illness. Outcome of treatment varies based on the severity of the disorder.

Complications

- Poor social skills
- Lack of interpersonal relation

4

TRAIT THEORIES

The trait approach to personality is one of the major theoretical areas in the study of personality. The trait theory suggests that individual personalities are composed broad dispositions. Consider how you would describe the personality of a close friend. Chances are that you would list a number of traits, such as outgoing, kind and even-tempered. A trait can be thought of as a relatively stable characteristic that causes individuals to behave in certain ways.

Unlike many other theories of personality, such as psychoanalytic or humanistic theories, the trait approach to personality is focused on differences between individuals. The combination and interaction of various traits forms a personality that is unique to each individual. Trait theory is focused on identifying and measuring these individual personalities' characteristics.

Gordon Allport's Trait Theory

In 1936, psychologist Gordon Allport found that one English-language dictionary alone contained more than 4,000 words describing different personality traits. He categorized these traits into three levels:

Cardinal Traits: Traits that dominate an individual's whole life, often to the point that the person becomes known specifically for these traits. People with such personalities often become so known for these traits that their names are often synonymous with these qualities. Consider the origin and meaning of the following descriptive terms: Freudian, Machiavellian, narcissism, Don Juan, Christ-like, etc. Allport suggested that cardinal traits are rare and tend to develop later in life.

Central Traits: These are the general characteristics that form the basic foundations of personality. These central traits, while not as dominating as cardinal traits, are the major characteristics you might use to describe another person. Terms such as intelligent, honest, shy and anxious are considered central traits.

Secondary Traits: These are the traits that are sometimes related to attitudes or preferences and often appear only in certain situations or under specific circumstances. Some examples would be getting anxious when speaking to a group or impatient while waiting in line.

Raymond Cattell's Sixteen Personality Factor Questionnaire

Trait theorist Raymond Cattell reduced the number of main personality traits from Allport's initial list of over 4,000 down to 171, 3 mostly by eliminating uncommon traits and combining common characteristics. Next, Cattell rated a large sample of individuals for these 171 different traits. Then, using a statistical technique known as factor analysis, he identified closely related terms and eventually reduced his list to just 16 key personality traits. According to Cattell, these 16 traits are the source of all human personality. He also developed one of the most widely used personality assessments known as the Sixteen Personality Factor Questionnaire.

Eysenck's three Dimensions of Personality

British psychologist Hans Eysenck developed a model of personality based upon just three universal trails:

Introversion/Extraversion

Introversion involves directing attention on inner experiences, while extraversion relates to focusing attention outward on other people and the environment. So, a person high in introversion might be quiet and reserved, while an individual high in extraversion might be sociable and outgoing.

Neuroticism/Emotional Stability

This dimension of Eysenck's trait theory is related to moodiness versus even-temperedness. Neuroticism refers to an individual's tendency to become upset or emotional, while stability refers to the tendency to remain emotionally constant.

Psychoticism

Later, after studying individuals suffering from mental illness, Eysenck added a personality dimension he called psychoticism to his trait theory. Individuals who are high on this trait tend to have difficulty dealing with reality and may be antisocial, hostile, non-empathetic and manipulative.4

Trait Theories of Personality

A trait is what we call a characteristic way in which an individual perceives feels, believes, or acts. When we casually describe someone, we are likely to use trait terms: I am, for example, somewhat of an introvert, a pretty nervous person, strongly attached to my family, frequently depressed, and awesomely intelligent. I have a good sense of humour, fond of languages, very fond of good food, not at all fond of exercise, and a little obsessive. You see: I have just given you ten traits that actually go a long way towards describing me!

Psychologists, especially personologists, are very interested in traits. They are especially interested in finding which traits are broad and possibly genetically based, as opposed to ones that are rather peculiar and can change easily. Over the years, we have had a number of theories that attempt to describe the key traits of human beings.

Carl Jung and the Myers-Briggs

One of the earliest trait theories was introduced by a colleague of Sigmund Freud's by the name of Carl Jung. Jung was never completely sold on Freud's ideas, and soon left his circle to develop his own theory. This is not the place to go into details, but one aspect of the theory concerned traits that Jung felt were inborn. These inborn, genetically determined traits are usually called temperaments.

Later, two students of Jung's theory named Myers and Briggs - mother and daughter - developed a personality test based on Jung's temperaments called the Myers-Briggs Type Inventory, or MBTI. It has gone on become the most famous personality test of all time.

The traits are seen as opposites, and the first set is introversion and extraversion. Introversion refers to a tendency to prefer the world inside oneself. The more obvious aspects of introversion are shyness, distaste for social functions, and a love of privacy.

Extraversion is the tendency to look to the outside world, especially people, for one's pleasures. Extraverts are usually outgoing and they enjoy social activities, but they don't like to be alone.

The majority of people in the world are extraverts, so introverts often feel a bit out of it. A society like ours is very pro-extravert, even to the point of seeing introversion as abnormal and shy people in need of therapy! There are some cultures, however, that see extraverts as the oddballs. We should note that it was Jung who first used the terms introversion and extraversion!

Jung believed that introversion-extraversion was either-or. You are born one or the other and remain that way for the rest of your life. Now you could, as an introvert, learn to behave more like an extravert, or, as an extravert, learn to behave more like an introvert. But you can't really switch. If this is true, that would suggest that introversion-extraversion is determined by a single gene, something that is pretty unusual even for more physical differences! Nevertheless, it seems that introversion-extraversion is a very significant and fairly stable trait.

Next, we have the contrast between sensing people and intuiting people. Sensing types, as the name implies, get all their information about life from their senses. They tend to be realistic, down-to-earth people, but they tend to see everything in rather simplistic, concrete, black-or-white terms.

Intuiting people tend to get their information from intuition. This means that they tend to be a little out of touch with the more solid aspects of reality - a little "flakey", you might say - but may see "the big picture" behind the details better. Intuiting people are often artistic and can be rather philosophical.

Again, the majority of people are sensing, and that can make intuiters feel rather lonely and under-appreciated. Our society tends to be distrustful of dreamers, artists, and intellectuals - but other societies may be more appreciative.

Next, there's the contrast between thinkers and feelers. Thinking people make their decisions on the basis of thinking - reasoning, logic, step-by-step problem solving. This works very well for physical problems, but can leave something to be desired when dealing with something as complex as people.

Feeling people make their decisions based on their feelings. This is not work while so well when trying to fix your car or your computer, feelings are a kind of intuition that works very well when dealing with people.

Half of all people are thinking and half are feeling, but the proportions differ when we start looking at gender: The majority of men are thinkers and the majority of women are feelers. This goes along well with old stereotypes as well as recent research: Men tend to do better with step-by-step problem solving, especially involving mechanical things; Women tend to do better in social situations. Some people have criticized Jung for this apparent sexism, but we should note that a good third of men are feelers, and a good third of women are thinkers, so it is not a simple "men vs. women" kind of thing. Plus, Jung said that there is no reason to value thinking over feeling - each has its strengths and weaknesses. Note also that feeling men may feel odd, as may think women. Stereotypes do the greatest harm when they prevent individuals from being what they in fact are!

The last contrast is judging versus perceiving. Judging people tend to be more like Freud's anal retentive types - neat, orderly, hardworking, always on time, scheduling things very carefully. College professors tend to be judging!

Perceiving people are more spontaneous. They prefer to do things as the spirit moves them. They are probably more fun than the judging types but, as you can imagine, they tend not to get things done. It often seems to us college professors that college students are all perceiving.

Actually, the distribution of judging and perceiving people is pretty even - 50-50.

When you take the Myers-Briggs or similar tests like the Keirsey, you get a set of four letters: I for introvert or E for extravert, S for sensing or N for intuiting, T for thinking or F for feeling, and J for judging or P for perceiving. I, for example, am an INFP, which is in fact quite accurate. My wife is an ISFJ -- she is more down-to-earth and organized than I will ever be. That's why she controls the family finances! On the other hand, we are both introverted and feeling, which means that you are more likely to find us crying over a rented movie than out at some wild party!

Hans Eysenck

Hans Eysenck was the first psychologist to make this trait or temperament business into something more mathematical: He gave long lists of adjectives to hundreds of thousands of people and used a special statistics called factor analysis to figure out what factors - trait dimensions - carry the most weight. He took the results of this work and created a test called the Eysenck Personality Questionnaire (EPQ).

Instead of making these traits either-or, like Jung did, he saw them as dimensions. His first trait dimension was, like Jung, extraversion-introversion. But rather than say you were one or the other, he gave you a score on extraversion-introversion: A low score meant you were introverted, a high score extraverted. Of course, this meant you could be halfway in-between - as in fact most people are!

His second trait dimension he called neuroticism. If you scored high on this scale, that meant you tended to be a very nervous, emotional sort of person. While it doesn't mean you are necessarily a neurotic, it does mean you are more likely to develop neurotic problems such as phobias, obsessions, compulsions, and depression than someone who scores low. Low neuroticism is nowadays often called emotional stability.

The third dimension is called psychoticism. He added this later in his research, after he had gotten more data from people who were in mental institutions. As the name implies, these are people with tendencies to psychosis, meaning that they are more likely to have problems dealing with reality. Psychotic people sometimes have hallucinations and often have delusions such as odd beliefs about being watched, perhaps by the CIA or even by creatures from other planets. A middle score on psychoticism might mean that you are a bit eccentric or that you take risks that other people aren't as likely to take. A low score means that you are pretty normal in this regard.

Eysenck's research gets a great deal of respect and most psychologists see his theory as on the right track.

Personality Traits

Personality traits are defined as the relatively enduring patterns of thoughts, Corresponding author`s feelings and behaviors that distinguish individuals from one another. Whether personality traits continue to develop in adulthood depends in part on how one defines relatively enduring. In the past, some researchers took that phrase to mean—and some empirical literature to indicate—that personality traits stopped changing in adulthood. Since 1994 cross-sectional and longitudinal studies of personality-trait change in adulthood have forced a re-evaluation of the assumption that personality traits do not change in adulthood.

Personality traits were measured by the Hindi version of the NEO Five-Factor Inventory (NEO-FFI), originally developed by Costa and McCrae. This inventory consists of sixty items and based on five-point Likert format.

The NEO-FFI was translated and validated into the Hindi language for use in the present research work. Item total correlation was performed to examine the validity of the item. Four items from neuroticism factor, five items from extraversion factor, and three items from agreeableness factor were excluded because of either low or negative items-total correlation. All twelve items of conscientiousness factor were retained; as they show satisfactory item total correlation. Items related to openness to change factor have not shown the satisfactory item total correlation so this factor was dropped in further analysis.

Procedure

In order to assess personality traits, subjects completed the NEO-FFI. Participants were asked to take their own time to complete the questionnaire. Completion of the questionnaire typically took less than 20 min.

Career Stage

Career theorists have proposed that under normal circumstances people will experience four career stages: exploration, establishment, maintenance and disengagement. Although common age ranges are associated with each stage, a variety of career, personal and life factors combine to determine the exact time when an individual moves from one stage to the next.

Most of the studies on Super's career stages development model have operationalized these stages by age. Gould and Slocum and Cron identified people as being in the trial stage if they are less than 30 years old, in the establishment stage if they are between 31 and 44, and in the maintenance stage if they are over 45.

Super's stages by the job tenure - amount of time an individual has been in a job. These researchers operationalized the trial stage as less than 2 years on the job, the establishment stage as between 3 and 10 years on the job, and the maintenance stage as over 10 years on the job. Consistent with previous studies we used multiple indicators to operationalize super's career stages.

By combining the developmental perspective of personality and Super's career stage theory the present study aimed to examine whether executives score differently on personality traits in their different career stages.

Review of Literature and Development of the Hypotheses

According to the five-factor theory, personality traits are "insulated from the direct effects of the environment" and are exclusively biological in origin. Regarding the change in personality traits the five-factor theory stated that: "Traits develop through childhood and reach mature form in adulthood; thereafter they are stable in cognitively intact individuals". More specifically, traits are said to reach maturity by the age of 30. The predicted stability is expected to last throughout middle age; though in old age personality could change again, being disrupted by cognitive decline. A commonly used metaphor for this pattern of change, based on a passage from William James, is that personality becomes "set like plaster" by the age of 30.

In its original formulation, the plaster hypothesis stated that changes in the Big Five traits after the age of 30 were nonexistent or trivial. Later, the authors of the five-factor theory indicated that the plaster hypothesis is "ripe for minor revision", as studies have shown changes in mean levels of personality traits after age 30. They interpret such changes as stemming from intrinsic biological maturation rather than social influences, and they still regard the plaster hypothesis as basically true: From age 18 to age 30 there are declines in Neuroticism, Extraversion, and Openness to Experience, and increases in Agreeableness and Conscientiousness; after the age of 30 the same trends are found, although the rate of change seems to decrease.

In contrast to a biological viewpoint, contextualist perspectives argue that traits are multiply determined, and that one important influence on traits is the individual's social environment. Contextualist perspectives thus predict plasticity: Change is complex and ongoing, owing to the many factors that can affect personality traits.

Cross-sectional and longitudinal research has converged on the findings that personality-trait development can and does occur in all age periods of adulthood, including old age. Specifically, cross-sectional research has shown that middle-aged individuals tend to score higher than young adults on agreeableness and conscientiousness and lower on extraversion, neuroticism and openness. Moreover, within middle age, 60-year-old participants scored higher than 40-year-old participants on most dimensions. Roberts and Mroczek also stated that personality traits continue to change, even in old age. One of the precepts of life-span orientation is that humans are open systems. That is, people retain the capacity to change at all ages. The changes in personality traits in middle and old age are by no means dramatic, but nonetheless they show that the life-span orientation applies to personality traits and that personality is not set like plaster at any point in the life course.

A recent literature review summarized previous studies of mean-level change on the Big Five. In this review, Roberts et al. Rationally categorized a wide variety of personality measures into the Big Five domains and summarized patterns of mean-level change that were consistent across studies. They concluded that, in general, Conscientiousness and Agreeableness tend to go up during adulthood, Neuroticism tends to go down, Openness shows mixed results across studies, and Extraversion shows no general pattern of change at the factor level. This basic pattern of findings has been reported in specific studies by researchers who argue that personality traits are affected by context as well as those who favour a strictly biological interpretation of traits The recent findings motivate a new generation of questions concerning why personality traits change more in young adulthood than in other periods of the life course and what the implications of the mostly positive trend in personality-trait change might be. Of course, one of the realities of any sweeping generalization is that it does not apply to all people. Much of this research needs to be replicated in non-Western cultures before firm conclusions are drawn. The present study is an effort in this regard to examine whether personality traits changes throughout the life span among managers in an Indian cultural context, which is different from Western culture.

On the basis of Super's career stage conceptualization and available research on personality development the following hypotheses were proposed: H1: Executives in their trial stage would score significantly differently on personality traits than the executives in the establishment and maintenance stages. H2: The difference of mean scores of personality traits between trial stage and establishment stage would be greater than the differences of mean scores between establishment stage and maintenance stage. On what feelings you want to train your body?

5

The Measurement of Personality

It should be pointed out that the various methods of assessing personality correspond closely to the basic personality theories we have just discussed. Projective tests such as the Rorschach test are closely linked to the psychoanalytic theory of personality.

Conversely, the various paper and pencil inventories are linked to traits theory. The methods outlined here are not perfect nor do psychologists agree on the meaning of the results of any particular test. There are three basic classes of measurement techniques. They are:

- Rating scales
- Unstructured projective tests
- Questionnaires

Rating scales

Characteristically, rating scales involve the use of a judge or judges who are asked to observe an individual in some situation. The judge employs the use of a checklist or scale that has been predesigned for maximum objectivity. Usually, if the checklist is used properly and the judges are well trained, the results can be fairly reliable and objective,

Typically, two types of situations are involved in personality assessment using rating scales. And these are,

- The Interview

- The observation of performance

In the interview, the judge asks the subject numerous open ended and specific questions designed to ascertain personality traits and general impressions. Generally, several interviews are necessary to gain impressions about underlying motives, if the interview is conducted properly, carefully, and systematically, the results can be reliable and valid. However, much depends upon the skill and sensibility of the person conducting the interview.

On the other hand the observation of a subject during some type of performance situation is the second kind of rating system used for ascertaining personality. As with the interview, observations can be effective if the checklist being used is well designed and planned, and if the observer is highly trained.

Projective Procedures

Projective procedures may also be used to identify traits, but they are commonly used to determine information about underlying motives. Projective techniques allow subjects to reveal their inner feelings and motives through unstructured tasks. These techniques are used in clinical psychology and are somewhat synonymous with the psychoanalytic and humanistic approaches to explaining personality. The underlying assumption is that if subjects perceive that there is no right or wrong responses, they will likely to be open and honest in their responses.

Several kinds of tests have been developed; amongst them are Rorschach test, the Thematic Apperception Test, the Sentence Completion Test, and the House-Tree-Person Test.

The Rorschach Test

Herman Rorschach, a Swiss psychiatrist, was the first to apply the inkblot to the study of personality. The Rorschach test was introduced in 1921, and remains the most famous of all the projective testing devices. The test material consists of ten cards. Each card has inkblot on it, which is symmetrical and intricate. Some of the cards are entirely in black and white, while others have a splash of colour or are nearly all in colour. The cards are prevented to the subject one at a time and in a prescribed order. As the cards are presented, the subject is encouraged to tell what he sees. Though the Rorschach test has not been used extensively by sport psychologists to evaluate personality in athletes but there is doubt that the responses the clients give to the Rorschach test contain a wealth of personal information. The question is whether the psychologist can accurately interpret the response. Research suggests that when the Rorschach is used as an objectively scored test, as recommended by Exner, it is a reliable and psychometrically sound test.

The Thematic Apperception Test

The Thematic Apperception Test, developed by Henry Murray and his associates in 1943 at the Harvard University Psychological Clinic has been used almost an extensively as the Rorschach test. The TAT is composed of nineteen cards containing pictures depicting vague situations, and one black card. The subject is encouraged to make up a story about each picture. In contrast to the vague blots in the Rorschach test, pictures in the TAT are rather clear and vivid. For example, the sex of the characters in the picture and their facial expressions are generally identifiable. It is believed that subjects reveal or project important aspects of their personalities they weave the characters and subjects in the picture into either an oral or a written story.

Like the first test that is the Rorschach test the TAT has not been used extensively by the sport psychologists to measure athlete`s personality. But its validity and reliability are highly dependent upon the skills and training of the individual administrating and interpreting the results.

Structured Questionnaire

The structured questionnaire is a paper-and-pencil test in which the subject answers specific true-false statements. There are many different kinds of questionnaire type personality inventories. For our purposes we will focus our discussion on the two most commonly used personality inventories. One of these inventories was developed to be used with individuals suffering from personality disorders, while the other was developed for normal populations.

1. Minnesota Multiphasic Personality Inventory

The Minnesota Multiphasic Personality Inventory (MMPI) is the most widely used of all personality inventories. It basically consists of a series of true/false questions designed to measure personality traits and clinical conditions of the athletes. The original version of the MMPI, composed of 550 items, was developed in the 1940s and is still in use. A revised version of the inventory, composed of 567 items, was developed in 1990 and named the MMPI-2. The authors also developed a new form of the inventory to be used with adolescents (MMPI-A). The traits measured by the MMPI-2 include the following hypochondria, depression, hysteria, psychopathic deviation, masculinity-feminity, paranoia, obsessive-compulsive behaviour, schizophrenia, hypomania, and social introversion. The inventory also includes items to detect lying and faking good/bad scores. While the MMPI-2 was designed specifically for use with clinical populations, it may be used with normal individuals.

2. Cattell`s Sixteen Factor Personality Inventory

Cattell`s Sixteen Factor Personality Inventory was developed by Robert Cattell, is based upon thirty-five personality traits originally identified by Cattell. Through a statistical process known as factor analysis, Cattell reduced the thirty-five specific traits to sixteen broader traits or factors. The current edition of the 16PF is titled the 16PF Fifth Edition and is composed of 195 items. The 16PF takes from thirty-five to fifty minutes to complete and is designed for adults, or those aged sixteen years and over. The 16PF measures 16 primary personality factors; it also included a social desirability index to assess faking good or bad.

6

The Six Stages of Modern Career Development

Career experts say that people will change careers (not jobs) five to seven times in a lifetime. That's why career management is an important life skill to develop and cultivate. The six stages of modern career development are:

- Assessment
- Investigation
- Preparation
- Commitment
- Retention
- Transition

Learning the characteristics of each stage will empower you to navigate through each stage easily and with more confidence.

In the Assessment Stage, you are getting ready for your life's work. This stage is characterized by unawareness, in that you are not sure what your values, strengths, and weaknesses are. You start to feel as though you want to know more about yourself and make a conscious effort to get in touch with who you really are.

Key Characteristics:

Taking assessment instruments

Working with a career counselor or career coach

In the Investigation Stage, you are researching what work exists in the world. This stage is characterized by feelings of confusion, in that you are not sure what career options exist for you. You may feel overwhelmed with all of the jobs and opportunities that exist as you begin the process of researching the modern world of work. But if you approach this stage with a positive frame of mind, you will find that you will learn about many possibilities you may have never considered.

Key Characteristics:

Researching the world of work

In the Preparation Stage, you are still getting ready to do your life's work. This stage is characterized by feelings of excitement, as you think of how wonderful it will be to perform meaningful work. However, there is still much work to be done, and to be successful, you have to prepare.

Key Characteristics:

Gaining knowledge and experience

Setting goals and adopting a success-oriented mind-set

In the Commitment Stage, you will feel confident that you have figured out what you are meant to do. Sometimes people have known all along what they were meant to do but could not commit to the process of making it happen, for whatever reason. At this stage, more than ever, you must focus your energy and keep your eye on the target.

Key Characteristics:

Conducting a job search

Negotiating and accepting a job offer

In the Retention Stage, you will feel comfortable in your career field, as you will now have figured out how things work in your industry. You will want to remain committed to your career by continually updating your skill set and staying current with industry standards.

Key Characteristics:

Providing first-class customer-service skills

Building a professional network

The Transition Stage is characterized by feelings of discomfort in that you are unsure of what you will be doing next. In this stage, you will learn to make conscious changes in your career direction.

Key Characteristics:

Making career changes

Developing resiliency

No matter what career stage you find yourself in now, you can be sure that you will enter and re-enter through these six stages many times throughout your lifetime.

Career Stages

The career stage approach is one way to look at career development. The issue of career stages is frequently based on Levinson's life stage development model. According to this model, people grow through specific stages separated by transition periods. At each stage a new and crucial activity and psychological adjustment may be completed. In this way, defined career stages can be, and usually are, based on chronological age. The age ranges assigned for each stage have varied considerably between empirical studies, but usually the early career stage is considered to range from the ages of 20 to 34 years, the mid-career from 35 to 50 years and the late career from 50 to 65 years.

According to Super's career development model the four career stages are based on the qualitatively different psychological task of each stage. They can be based either on age or on organizational, positional or professional tenure. The same people can recycle several times through these stages in their work career. According to the Career Concerns Inventory Adult Form, the actual career stage can be defined at an individual or group level. This instrument assesses an individual's awareness of and concerns with various tasks of career development. When tenure measures are used, the first two years are seen as a trial period. The establishment period from two to ten years means career advancement and growth. After ten years comes the maintenance period, which means holding on to the accomplishments achieved. The decline stage implies the development of one's self-image independently of one's career.

Because the theoretical bases of the definition of the career stages and the sorts of measure used in practice, it is apparent that the results concerning the health- and job-relatedness of career development vary, too.

Career Stage as a Moderator of Work-Related Health and Well-Being

Most of career stage as a moderator between job characteristics and the health or well-being of employees deal with organizational commitment and its relation to job satisfaction or to behavioural outcomes such as performance, turnover and absenteeism. The relationship between job characteristics and strain has also been studied. The moderating effect of career stage means statistically that the average correlation between measures of job characteristics and well-being varies from one career stage to another.

Work commitment usually increases from early career stages to later stages, although among salaried male professionals, job involvement was found to be lowest in the middle stage. In the early career stage, employees had a stronger need to leave the organization and to be relocated. Among hospital staff, nurses' measures of well-being were most strongly associated with career and affective-organizational commitment. There happens emotional attachment to the organization. Continuance commitment and normative commitment increased with career stage.

A meta-analysis was carried out of 41 samples dealing with the relationship between organizational commitment and outcomes indicating well-being. The samples were divided into different career stage groups according to two measures of career stage: age and tenure. Age as a career stage indicator significantly affected turnover and turnover intentions, while organizational tenure was related to job performance and absenteeism. Low organizational commitment was related to high turnover, especially in the early career stage; whereas low organizational commitment was related to high absenteeism and low job performance in the late career stage.

The relationship between work attitudes, for instance job satisfaction and work behaviour, has been found to be moderated by career stage to a considerable degree. Among employees of public agencies, career stage measured with reference to organizational tenure was found to moderate the relationship between job satisfaction and job performance. Their relation was strongest in the first career stage. This was supported also in a study among sales personnel. Among academic teachers, the relationship between satisfaction and performance was found to be negative during the first two years of tenure.

Career stage has dealt with men. Even in which the sex of the respondents was not reported, it is apparent that most of the subjects were men. Somebody tested how the career stage models of Levinson and Super described differences in the career attitudes and intentions among professional women. The results suggest that career stages based on age were related to organizational commitment, intention to leave the organization and a desire for promotion. These findings were, in general, similar to the ones found among men. However, no support was derived for the predictive value of career stages as defined on a psychological basis.

Stress have generally either ignored age, and consequently career stage, designs or treated it as a confounding factor and controlled its effects. Few of the psychologists contrasted the effects of stress in mid-career to its effects in early and late career using age as a basis for their grouping of workers. Perceived ill health was not related to job stressors in mid-career, but work pressure and underutilization of skills predicted it in early and late career. Work pressure was related also to somatic complaints in the early and late career group. Underutilization of abilities was more strongly related to job satisfaction and somatic complaints among mid-career workers. Social support had more influence on mental health than physical health, and this effect is more pronounced in mid-career than in early or late career stages.

When adult male and female workers were grouped according to age, the older workers more frequently reported overload and responsibility as stressors at work, whereas the younger workers cited insufficiency, boundary-spanning roles and physical environment stressors. The older workers reported fewer of all kinds of strain symptoms: one reason for this may be that older people used more rational-cognitive, self-care and recreational coping skills, evidently learned during their careers, but selection that is based on symptoms during one's career may also explain these differences. Alternativel) it might reflect some self-selection, when people leave jobs that stress them excessively over time.

Among managers, the relationship between job demands and control on the one hand, and psychosomatic symptoms on the other, was found to vary according to career stage. Among some managers, job demands and control had a significant effect on symptom reporting in the middle career stage, but not in the early and late stage, while among other managers, the long weekly working hours and low job control increased stress symptoms in the early career stage, but not in the later stages. The managers, being in the construction trades, had high workloads already in their early career stage, whereas on some other managers-these were public sector workers - had the highest workloads in their middle career stage.

To sum up the results on the moderating effects of career stage: early career stage means low organizational commitment related to turnover as well as job stressors related to perceived ill health and somatic complaints. In mid-career the results are conflicting: sometimes job satisfaction and performance are positively related, sometimes negatively. In mid-career, job demands and low control are related to frequent symptom reporting among some occupational groups. In late career, organizational commitment is correlated to low absenteeism and good performance. Findings on relations between job stressors and strain are inconsistent for the late career stage. There are some indications that more effective coping decreases work-related strain symptoms in late career.

Interventions

Practical interventions to help people to cope better with the specific demands of each career stage would be beneficial. Vocational counselling at the entry stage of one's work life would be especially useful. Interventions for minimizing the negative impact of career plateauing are suggested because this can be either a time of frustration or an opportunity to face new challenges or to reappraise one's life goals. Results of age-based health examinations in occupational health services have shown that job-related problems lowering working ability gradually increase and qualitatively change with age. In early and mid-career they are related to coping with work overload, but in later middle and late career they are gradually accompanied by declining psychological condition and physical health, facts that indicate the importance of early institutional intervention at an individual level. Both in research and in practical interventions, mobility and turnover pattern should be taken into account, as well as the role played by one's occupation in one's career development.

The 5 Career Stages

We may always ask our clients their age. We don't mean to make them feel bad about how old they are. But someone that's 55 is at a different stage in their career and life than someone that's 25 or 45. One of the pioneers in the field of career development, Dr. Donald Super argued that people pass through five career stages during their life span.

Within each stage mastery of certain tasks allows people to function successfully within that stage while preparing them to move on to the next task. Each stage is loosely affiliated with a chronological period and is characterized by work attitudes, behaviors, types of relationships, and the aspects of work that are valued.

1. **Growth–** The early years (4 to 13 years old) is a time when the individual first becomes aware of the future. People start to find ways to develop competencies and to achieve in order to increase control over their life.
2. **Exploration–** from the early teens to mid-twenties, people begin to crystallize, specifies and implements an occupational choice. Different roles are tried and various occupational options are explored though school, leisure,

part-time work and volunteering. "Trial jobs" may be tested before more firmly finding a more stable and appropriate fit.

3. **Establishment**– in the mid-twenties through mid-forties, typically a suitable field is selected and efforts are made to secure a long-term place in the chosen career. Young adulthood tends to be a time for stabilizing, consolidating, building momentum and moving up. Obtaining certifications, credentials, and advanced degrees may be the norm.

4. **Maintenance**- This stage usually happens in the mid-forties to mid-sixties and is characterized by constancy:

   1) Holding on, or

   2) Keeping up.

      Continuity, stress, safety and stability tend to be the standard. Sometimes people feel risk adverse with various career options which may lead to frustration or even depression. In middle adulthood we may ask ourselves, For men, state of health or career accomplishment may predominate. Women sometimes perceive this period as an opportunity to pursue new personal or professional goals now that their nurturing role has peaked.

5. **Disengagement**– The mid-sixties is typically marked by decelerating from formal employment to finding new roles with a view to retirement. Baby Boomers are teaching us that this stage should be more appropriately named "Re-inventment." They are completely redesigning the notion of "retirement" preferring to work in some form while pursuing new or renewed outside interests. In later adulthood, there may be a need to assist or mentor younger members of society or seek self-employment.

   However, it has limitation due to the rapidly changing nature of work and each person's own circumstances. Not everyone transitions through these five stages at fixed ages or in the same manner. I have learned in my private practice it is more common nowadays to move back and forth more frequently from the Exploration to Reinventment stages.

For example, before entering the Maintenance stage, many people are asking the mid-life question, "Do I want to do this job for the next twenty years?" Eventually, they decide to either:

1) Hang on and enter the Maintenance stage, or
2) Let go and change their job, company, or career and then recycle back to an earlier stage in order to move in a new direction. For others, their career is without boundaries based on skills and abilities that function independent of a set timeline.

Your self-image evolves throughout your life as a result of experience. You successively refine your uniqueness over time and make adaptation in your career choices. As we go through life stages our priorities change. Career ideas that you had at 25 might not be relevant when you are 45 or 55.

So, what career stage do you actually find yourself in chronological terms? What specific stage do you think you **need** to be in? Perhaps this is the time to step back and reflect on where you are. Maybe you need to return to the Exploration stage and re-evaluate your skills, values, interests, personality traits and core priorities. At your stage, perhaps you what to find out what else is out there and you want to begin a process of career discovery. What are the financial and time considerations of the choices you are making at your stage? What types of planning and preparation needs to be done at each stage both personally and professionally? Can you list three to five specific issues that you need to address right now? What are some issues you'll face as you pass through future stages?

These are all important questions that you will need to get clarity on in order to have successful and satisfying career development. Just knowing that as we age, we will be progressing through career stages, can be insightful, freeing and can have a profound effect on our professional development.

Remember: It's never too late to reinvent yourself!

Tom' Tip: "Change is the only constant." – Anonymous

Measures Career Stages

Smart stated that age and tenure both are used as a measure of Super's career stage in previous studies. But this has often led to confusing and contradictory results. Therefore, in the present study, career stage was measured in term of both executives' age and organization tenure simultaneously to provide direct comparison between the two variables and avoid this previously identified limitation.

Executives' age was divided into three time-frame subgroups: (a) up to 30 years, (b) 31 to 44 years, (c) older than 45 years. Similar to previous operationalizations;, the first subgroup represents the trial career stage, the second represents the establishment or settling-down stage, the third represents the maintenance stage.

Organizational tenure was divided into three subgroups: (a) less than 3 years, (b) 3 to 10 years, and (c) 11 years or more. Again, similar to previous operationalizations, the first tenure subgroup represents the trial stage; the second, the establishment stage; and the third, the maintenance stage.

7

Personality Conflict

Personality conflicts are perhaps one of the most challenging areas of corrective action. It is not uncommon to have employees who simply don't get along well with one another or to have one employee who doesn't seem able to get along with anyone, including yourself. It is critical to understand that you cannot discipline or terminate someone simply because of a "bad attitude" or "poor personality." Define the personality or attitude in terms of behavior and address the behaviors that need improvement. All courses of action that you take need to be focused on behavioral objectives. The following suggestions explain how to maintain that necessary focus.

Suggested Actions

- Define the personality problem as a behavioral problem that is affecting job performance:

  Change: "I don't like your attitude and your personality isn't a good fit for here."

  To: "The language you used in yesterday's meeting was inappropriate. It is also inappropriate to roll your eyes at other staff members. These behaviors prevent us from openly discussing things in team meetings which is affecting our performance."

- After defining the problem as a specific inappropriate behavior or series of inappropriate behaviors, follow the suggestions set out in the inappropriate behavior section.

  If the inappropriate behavior still exists after you have

redefined the situation as a specific behavior issue and have taken the actions outlined in the inappropriate behavior section then begin the progressive discipline process.

When is a personality conflict, not a personality conflict?

Could you tell me exactly what a "personality conflict" is? How does one differentiate a personality conflict from let's say emotional abuse or a personality disorder?

Answer: Here's a little quiz for you: What is the most commonly misapplied diagnosis when it comes to workplace conflicts? You guessed it: a personality conflict. I suspect the reason we like to attribute so many circumstances to this generic term is because we haven't dug deep enough to determine the cause of the problem and it's an easy phrase to slap on just about anything.

For example, two women who share the same office are not speaking. The manager has tried to intervene to no avail. He thinks they have a personality conflict but if you dive deeper, you will find that one of them always takes a long lunch and comes in late. But because the boss doesn't want to lose her, he refuses to say anything about it. Meanwhile the diligent employee has started to make not-so-subtle remarks about her coworker's work habits. I think not. Personality has nothing to do with it. If the manager had stepped in and done the right thing, this issue would never have occurred.

Then there is the case of the two professionals who worked for a small company run by an entrepreneur. The owner hated bureaucracy and avoided policies because he felt they would turn his company into the kind of big corporation he came from. The problem was that he made decisions about benefits and "policies" in casual conversations with individual employees. He never wrote anything down —no handbook, no memos, and no emails. When the two professionals discovered that each of them had different "agreements" with the owner, conflict erupted. He called it a petty personality conflict. It was far from it.

Now let's take the case of the young, single woman who is always discussing her private affairs with the rest of the office. Every day is a soap opera. The rest of the female employees are married and start to feel angry and offended when the young woman begins to share the explicit details of her affair with a married man. The parties involved have different values and have made drastically different life choices. Their conflict has nothing at all to do with work.

Now let's tread into murkier waters of emotion abuse. Since I'm not a psychiatrist or a judge, I can only give you my opinion. To me, emotional abuse at work is any behavior that is demeaning, threatening, disrespectful, belittling and is intended to destroy self-esteem. For example, much like an emotionally abusive spouse, an abusive boss can be a screamer, silent manipulator or verbal abuse.

A personality disorder can take many forms. For instance, en employee may have an obsessive-compulsive disorder, where he or she checks and rechecks every last detail of everything he or she does. Left untreated, these individuals usually spiral down and are unable to work, for fear of leaving the house with a toaster plugged in or the hose running. They have to keep checking, even though they just checked a minute ago.

Another example would be an excessively paranoid individual who believes that management is always out to get him, in spite of his manager's repeated attempts to patiently work through the employee's concerns. In both cases, these folks need professional help. No matter how sympathetic their managers are, they won't be able to fix their employee's problems.

There is no doubt that there are many different kinds of conflicts at work. And most workplaces reflect the whole spectrum of human behavior. The secret is in knowing what the core cause is and that makes all the difference in what you call it and how you fix it.

Personality Psychology

Personality psychology is the branch of psychology that strives to understand the traits and tendencies that make us unique.

What is Team Conflict?

Conflict in teams is a significant contributor to poor performance, but seldom is it dealt with effectively. This section will help with a basic understanding of conflict and how to deal with it.

Conflict Vs. Disagreement

First, let's distinguish between conflict and disagreement, as there are common misconceptions.

Disagreement adds needed argumentative dialogue to team discussions. Team members possess a cross-section of life experiences and general diversity that add richness to any discussion. Disagreement is inevitable and healthy to these discussions and, in fact, is needed to produce meaningful results. Disagreement should be welcomed by all team members and understood as a valuable contribution.

Conflict is disagreement that has elevated to an emotional level. Conflict interrupts the normal work activity by distracting team members, undermining team spirit and, ultimately, loyalty. This is not the kind of environment that allows a team to perform at its best.

What is Conflict?

"Conflict" is a word that causes most of us a great degree of discomfort, anger, frustration, sadness, and pain.

The dictionary defines "conflict" as "a struggle to resist or overcome; contest of opposing forces or powers; strife; battle. This is a state or condition of opposition; antagonism; discord.

A painful tension set up by a clash between opposed and contradictory impulses." No matter how hard we try to avoid it, conflict periodically enters our lives.

In the workplace, a simple disagreement between team members, if unresolved, may escalate into avoidance, inability to work together, verbal assaults, and resentment. In the worst cases, it may also lead to hostility and eventual separation from the organization. Therefore, it is important that the conflict be resolved as soon as possible.

Causes of Conflict

There is miscommunication leading to unclear expectations

How often do we give instructions to someone, only to have those instructions misinterpreted? The ability to communicate is one of our most commonly used skills. As such, we sometimes take it for granted so that the words we use to communicate don`t always state the picture in our minds. When this occurs, errors often result that lead to frustration. Depending on a multitude of factors (stress level for one), the error sometimes results in conflict if neither person is willing to accept responsibility for it.

Example: Miscommunication

Personality Clashes

We are all different. Experts say that our personalities are genetically determined resulting in different sets of preferred behaviours. Please see if the following comparisons ring some bells for you:

Some people are	Where others are
Outgoing Spontaneous, and Talkative.	introspective, serious, and quiet
Intuitive...shoot from the hip	Detailed...evaluate, ponder, and consider
Feeling and emotional	Logical and analytical
Concerned for people	Concerned for concepts
Structured, ordered and planned	Flexible, go with the flow, unplanned

These natural sets of differences are some of our greatest strengths as individuals and teams; however, they are also sources of conflict. If I, for example, prefer to look at only the "big picture," then I may become frustrated by your attempts to discuss details. You, on the other hand, may see me as irresponsible for not doing the analysis. Result: potential conflict.

Causes of Conflict:-

Differences in values and beliefs

There are differences in acquired values.

From the moment we are born, we begin acquiring our values system. Our values are the beliefs we hold that help us to make decisions about what is right or wrong, good or bad, and normal or not normal. Our values come from parents, siblings, friends, mentors, coaches, teachers, books, churches, movies, television, and music...life in general. No two people ever have the same life experience, so we ultimately have different sets of values and beliefs that guide our decisions and behaviour.

People struggle over religion, politics, race, humanitarian issues, ethics and morals, abortion, sex, and more. In extreme cases, some people will, literally, die for their beliefs. So this "gut level" value system is a strong driver of behaviour and a frequent source of conflict in our lives and in our teams.

Example: Differences in values and beliefs.

Causes of conflict– There is underlying stress and tension

Our lives today place enormous demands on our time and energy. But frequently those demands exceed our capacity to deal with them. Never the less, we come to work and attempt to function normally with our team members. Too often, however, this underlying stress surfaces at the slightest provocation, and we find ourselves in conflict. Dr. Wayne Dyer uses the analogy of an orange to describe this:

When you squeeze an orange, you get orange juice. Why? Because that`s what is inside. When you squeeze (metaphorically) a person, you also get what the person is holding inside. One of our greatest lessons is to understand that a person`s angered response to us may have nothing to do with us at all. They may simply be reflecting other stresses in their lives. Knowing this makes it easier to respond in a more tempered, appropriate, and responsible manner, if we don`t understand this important principle. We may react to their anger with similar anger, elevating to one of conflict.

Causes of Conflict– Ego Problems

Ego is another strong driver of our behaviour and decisions. Ego wants us to be "right" and moves us into defending our position, sometimes unreasonably.

One of the quickest ways to diffuse an argument or conflict is to admit one`s mistakes: At a minimum, move out of ego and attempt to see the situation from the other person`s point of view. There is a saying that it takes a big person to admit his mistakes. Consider this the next time you are defending yourself, and not sure why.

Resolving Conflicts

Key Principles in Resolving Conflict

- Take shared responsibility for the conflict;
- Recognize and appreciate differences among people;
- Preserve individual dignity;
- Listen carefully and with empathy, listen to understand, communicate, don`t debate;
- Be calm... don`t give in to emotional outbursts or reactions;
- Vulnerability is a key to successful resolution, therefore open up and share your feelings;
- Don`t assume people are being difficult internationally;
- Choose a safe place or person with whom you can vent and clearly the issues for yourself;
- Generate solutions.... Find agreement;
- Follow-up to assure resolution and modify as necessary;

The 7 Steps to Conflict Resolution

To Resolve a Conflict	What to say or do?	Why?
Calm yourself	Take a deep breath, say "relax"	Clears thinking, models control
Restore order	Take a "Time Out"	Stops the fight, contains the damage
Hear their stories	"Help me understand your concern."	Gathers information, defuses tension
Listen carefully	Eye contact, don't interrupt	Honors the need to be heard

Generate solutions	"How could we resolve this?"	Moves from accusations to solutions
Agree on a solution	"Would this work for you?"	Moves to resolution, brings closure
Test for satisfaction	"Are you sure this will work for you?"	Assures clear communication

Key Points to Remember

- Be a model of calm and control
- Don't give in to emotional outbursts
- Don't assume people are being difficult intentionally
- Find a quiet place in the store to resolve conflicts....privately
- Set some ground rules for the discussion:
- No raising of voices
- This is not a debate
- Speak only for yourself..."I" phrases
- Confront the issues, not the people
- Maintain or enhance self-esteem

Suggestions - How to deal with team conflict

How to minimize team conflict in the first place

- How to keep disagreement from escalating into conflict
- How to deal with difficult team members

Suggestions - How to minimize team conflict in the first place

Get training on the human dynamics of working together. Most teams are first taught how to do the work the task at hand. Secondarily (but usually never) they are taught the human dynamics of working together. As a result, when issues arise stemming from interpersonal relationships, team members may find themselves frustrated, in frequent disagreements, and conflict simply because they really don't understand each others' unique personalities, beliefs, and eccentricities.

As a first step in developing your new team, find someone who can teach team dynamics. Such training should include:

- Understanding Personality
- Personal Values and Beliefs
- Active Listening
- Stages of Group Development
- Consensus Decision Making
- Personal and Team Motivation
- Conflict Resolution

As a beginning, these training elements will help team members to understand the causes of disagreement and conflict, the basics of human behavior, and how to appreciate their differences rather than resist them. Much conflict can be avoided or minimized with knowledge of who we are.

Note: When we ask existing teams the question "If you had it to do over again, would you do anything differently in your development?" they usually tell us they would get training earlier on how to work together better, that is, the human side of team operation.

Suggestions - How to minimize team conflict in the first place

Develop a conflict resolution process.

One of the simplest, but best, ways to deal with team conflict is to develop a resolution process at the outset. Talk openly with team members about the inevitability of conflict and how they would like to deal with it. The process can vary from team to team, so have them develop their own process that each member agrees to abide by. When conflicts arise, use the process.

Suggestions - How to keep disagreements from escalating into conflict

Members should openly discuss the importance of disagreement. Without disagreement, teams can fall into groupthink, or a tendency to blindly agree on issues without proper questioning. Team members sometimes fall into this trap to avoid conflict, but then fall into a larger one of making poor decisions. So disagreement and good argumentative discussion should be encouraged.

However, team members should also discuss when this disagreement crosses the line into open conflict that may be harmful to the team.

Are there specific behaviors that fan the flames of conflict?

Are there issues that should be discussed that team members have been reluctant to address? Better to address them now than wait until they reach the boiling point.

Having a discussion about disagreement and conflict is an adult behavior. And team members will appreciate the opportunity to have this discussion and resolve in advance what could be a barrier to higher team performance.

Suggestions - How to deal with difficult team members

Difficult people don't want to be difficult; however, sometimes there are underlying factors that may be driving this behavior such as:

- Unpleasant childhood experiences
- Current stress in the family! Or another situation
- Significant personality differences!
- A general lack of motivation

Regardless of what is provoking this behavior, someone needs to approach the person and help him understand the impact the behavior is having on the team. When someone is in a "bad place," however, this kind of confrontation is usually met with resistance, or worse, ambivalence. The person just simply may not care.

The first step to approaching this person is to develop a relationship or friendship that provides safety and security. This requires one to completely detach from the emotions associated with the difficult person (Example).

As this relationship develops, the person will become more receptive to feedback and more likely to experience a "turnaround" in their behavior.

This approach requires us to move to a higher plane of thinking, one that understands that more can be accomplished through unselfish service to others rather than retreating to the comfort of our own egos and insecurities.

8

Psychological Theories of Personality Development

Personality development has been a major topic of interest for some of the most prominent thinkers in psychology. Our personalities make us unique, but how does personality develop? How exactly do we become who we are today?

In order to answer this question, many prominent theorists developed theories to describe various steps and stages that occur on the road of personality development. The following theories focus on various aspects of personality development, including cognitive, social and moral development.

Piaget's Stages of Cognitive Development

Jean Piaget's theory of cognitive development remains one of the most frequently cited in psychology, despite being subject to considerable criticism. While many aspects of his theory have not stood the test of time, the central idea remains important today: children think differently than adults. Learn more about Piaget's groundbreaking theory and the important contributions it made to our understanding of personality development.

Freud's Stages of Psychosexual Development

In addition to being one of the best-known thinkers in the area of personality development, Sigmund Freud remains one of the most controversial. In his well-known stage theory of psychosexual development, Freud suggested that personality develops in stages that are related to specific erogenous zones. Failure to successfully complete these stages, he suggested, would lead to personality problems in adulthood.

Freud's Structural Model of Personality

Freud's concept of the id, ego and superego has gained prominence in popular culture, despite a lack of support and considerable skepticism from many researchers. According to Freud, three elements of personality—known as the id, the ego, and the superego, they work together to create complex human behaviors

Erikson's Stages of Psychosocial Development

Erik Erikson's eight-stage theory of human development is one of the best known theories in psychology. While the theory builds on Freud's stages of psychosexual development, Erikson chose to focus on how social relationships impact personality development. The theory also extends beyond childhood to look at development across the entire lifespan.

Kohlberg's Stages of Moral Development

Lawrence Kohlberg developed a theory of personality development that focused on the growth of moral thought. Building on a two-stage process proposed by Piaget, Kohlberg expanded the theory to include six different stages. While the theory has been criticized for a number of different reasons, including the possibility that it does not accommodate different genders and cultures equally, Kohlberg's theory remains important in our understanding of how personality develops.

Personal Development

Personal development includes activities that improve awareness and identity, develop talents and potential, build human capital and facilitates employability, enhance quality of life and contribute to the realization of dreams and aspirations. The concept is not limited to self-help but includes formal and informal activities for developing others, in roles such as teacher, guide, counselor, manager, coach, or mentor. Finally, as personal development takes place in the context of institutions, it refers to the methods, programs, tools, techniques, and assessment systems that support human development at the individual level in organizations.

At the level of the individual, personal development includes the following activities:

- Improving self-awareness;
- Improving self-knowledge;
- Building or renewing identity;
- Developing strengths or talents;
- Improving wealth;
- Spiritual development;
- Identifying or improving potential;
- Building employability or human capital;
- Enhancing lifestyle or the quality of life;
- Improving health;
- Fulfilling aspirations;
- Initiating a life enterprise or personal autonomy;
- Defining and executing personal development plans;

Personal development in higher education

Personal development has been at the heart of education. Some people emphasize personal development as a part of higher education. They made a statement interpretable as referring to personal development: if there is one thing more than another which absolutely requires free activity on the part of the individual, it is precisely education, whose object it is to develop the individual.

Arthur Checkering defined seven vectors of personal development for young adults during their undergraduate years:

- Developing competence;
- Managing emotions;
- Achieving autonomy and interdependence;
- Developing mature interpersonal relationships;
- Establishing identity;
- Developing purpose;
- Developing integrity

Personal development took a central place when the Dearing Report declared that universities should go beyond academic teaching to provide students with personal development. A Quality Assessment Agency produced guidelines for universities to enhance personal development as:

A structured and supported process undertaken by an individual to reflect upon their own learning, performance and achievement and to plan for their personal, educational and career development;

Objectives' related explicitly to student development; to improve the capacity of students to understand what and how they are learning, and to review, plan and take responsibility for their learning

Personal development in the workplace

Abraham Maslow proposed a hierarchy of needs with self actualization at the top, defined as:

The desire to become more and more what one is, to become everything that one is capable of becoming.

Since Maslow himself believed that only a small minority of people self-actualize the estimated one percent — his hierarchy of needs had the consequence that organizations came to regard self-actualization or personal development as occurring at the top of the organizational pyramid, while job security and good working conditions would fulfill the needs of the mass of employees.

As organizations and labor markets became more global, responsibility for development shifted from the company to the individual. Management thinker Peter Ducker wrote in the Harvard Business Review:

We live in an age of unprecedented opportunity: if you've got ambition and smarts, you can rise to the top of your chosen profession, regardless of where you started out. But with opportunity comes responsibility. Companies today aren't managing their employees' careers; knowledge workers must, effectively, be their own chief executive officers. It's up to you to carve out your place, to know when to change course, and to keep yourself engaged and productive during a work life that may span some 50 years.

Companies must manage people individually and establish a new work contract. On the one hand the company must allegedly recognize that personal development creates economic value: "market performance flows not from the omnipotent wisdom of top managers but from the initiative, creativity and skills of all employees".

On the other hand, employees should recognize that their work includes personal development and "embrace the invigorating force of continuous learning and personal development".

It had corresponded to a change in career development from a system of predefined paths defined by companies, to a strategy defined by the individual and matched to the needs of organizations in an open landscape of possibilities. Another contribution to the career development came with the recognition that women's careers show specific personal needs and different development paths from men. The women's careers had a major impact on the way companies view careers. Further, the career as a personal development process came from study on the relationship with career change and identity change, indicating that priorities of work and lifestyle continually develop through life.

Personal development programs in companies fall into two categories: the provision of employee benefits and the fostering of development strategies.

Employee benefits have the purpose of improving satisfaction, motivation and loyalty. Employee surveys may help organizations find out personal-development needs, preferences and problems, and they use the results to design benefits programs. Typical programs in this category include:

- Work-life balance;
- Time management;
- Stress management;
- Health programs;
- Counseling;

Many such programs resemble programs that some employees might conceivably pay for themselves outside work: yoga, sports, martial arts, money-management, positive psychology, etc.

Social relation: for personal development

In general term social relation is the relationship between two or more individuals which form the basis of social group or social structure.

Everybody needs somebody and this is why social relation is typically used to refer to circle of friends and connection you have, the relationships with them and the quantity and quality of the time you spend with them. There are huge numbers of people who do not care about social relation which probably affects their mood, self-image, and performance in other areas of life. Whether they like it or not but the social behavior and interactions tend to matter a lot more than how much money you make or how successful you are in your career.

There are so many advantages of social relation like:

Fun: social relation helps to get rid of our boring life as some of us have boring friends, which we don't really want to connect with and this is one of the reasons why we lose interest in social relationship. The problem is not the social life and what social life offer but the type of social relation we form. When you will have circle of friends who appreciate you and vice-versa then the social interactions becomes highly enjoyable. This is how interaction with other adds fun and joy to your life.

Fulfillment: All human being are social creatures and all have innate need to interact with other people, and to build social relationships. Research shows that strong, quality relationships whether it is small or big is one of the most visible contributors to happiness and fulfillment.

Growth: The theory of growth is just not developing yourself but developing the social relationship. The overall growth and success are influenced to a large extent by the people in your social circle. The more positive thinking people you will have in your circle that aim high and always have something valuable to teach you then the speed of your growth will be high. This will ultimately boost your personal development, your career.

So in short we can conclude that building social relation can be very helpful to nourish your personal development skills.

Essential Life Skills for Personal Development

Personal development is the pursuit of developing, honing and mastering the skills that help us become the best that we can, with all that we have. It is the reaching for, and realizing of, our full potential as human beings.

We all want to live full; productive lives, but sometimes, we just don't know where to begin. There is so much information 'out there' that it can be overwhelming and hard to sort out. Depending on the problem, what seems to work for one person, may not necessarily work for everyone. There are so many different programs, strategies and techniques that it's hard to chose the right one.

One thing, however, is certain. If we want to accomplish anything in life and realize our full potential, we must have some skills - in these case life skills.

Where do you begin?

You begin by establishing a firm foundation. That foundation is "you". You must know who you are, what you want, and what you are capable of. You must then determine which values, goals and principles you will set up to guide your actions.

Learning about and applying the Essential Life Skills will help you to:

- Know and understand yourself better;
- Live life more consciously and deliberately;
- Attain personal satisfaction and fulfillment.

Often, the hardest part in any endeavor is getting started, however once you do, there is a surprising snowball effect. You will begin to feel good about what you're doing and you'll want to continue. You will want to keep improving yourself and you'll want to become the best that you can be.

As you continue on the journey of personal development, you will become aware that there is so much more knowledge and information to be discovered, and uncovered, than you ever thought possible - knowledge about yourself, knowledge about others, knowledge about life and the world around you.

According to Aristotle: - *"All men by nature desire Knowledge"*

Fourteen skills for Personal Development

It's helpful to take a high level perspective on personal development, looking broadly at all the important skills. This can help identify areas you may be weak in, and serve as a mental checklist to help you focus. These are some of the most important.

1. **Continuous Learning:** Take responsibility for your own lifelong personal development, knowing that it's the secret to success in all areas of life, and your most valuable asset. Your knowledge and experience is personal wealth you can never lose.
2. **Positivity:** Find the positive aspect in everything, accepting what cannot be changed, and acting on what can, towards a positive result. Don't let circumstances drive your state of being, let your state of being drive your results. Smile and use humor to help drive a positive attitude and share it with others.
3. **Personal Excellence:** Strive to do your best in everything you do. If it's worth doing, it's worth doing well. Not only to achieve superior results, but so that if you fail you can remain positive rather than wishing you had tried harder. Look to be creative in your personal excellence rather than follow the safest, lowest risk path.
4. **Honesty:** Maintain honesty and integrity in all that you do. Trust and respect by others is hard earned, yet can be permanently lost over the smallest lie. Decide that honesty is one of your core values, and stick to it at all costs.

5. **Know Yourself:** You are a unique individual with your own strengths and weaknesses, all others have these also. The difference with a successful person is they have learned to leverage their strengths, and to identify their weaknesses so they can improve those areas through learning and practice. Accept yourself as you are without comparison to others.
6. **Appreciate the Moment:** Seek happiness in the present moment, even in the simple things around you such as nature, a good conversation. Excessive contemplation on the past or future steals this appreciation from you. Some techniques to support this are getting out in nature, and meditation.
7. **Communicate Well:** Learning to communicate well with others is an essential skill, which includes speaking, writing, listening well, and body language. Nearly everyone can still find areas to grow with these skills, and should receive constant attention. Becoming a good listener is one of the top areas of improvement you should seek, become truly interested in other people and it will work wonders.
8. **Embrace Mistakes:** Mistakes are a prerequisite to personal development; you must choose either courage and success, or safety and being average. Mistakes are a learning opportunity so long as you are not repeating your mistakes, if you have learned from a failure then pat yourself on the back. Carry this to other's mistakes also; forgive them with this understanding in mind.
9. **Embrace Change:** Change is a constant part of life, and integral to personal development and success. Become excited about change, and look to it with a positive attitude. Be a leader by being the initiator of positive change. If you fear change, then you fear personal development, as self improvement is about changing yourself.
10. **Increase Your Value:** You should know that personal development is not a selfish activity; rather it increases your

value to others, ranging from relationships, business, and career. Too many people focus on changing others, which rarely works as well as changing yourself. Make a difference externally by changing yourself internally.

11. **Be Giving:** Seek to be giving, compassionate, loving, yet humble, knowing you receive what you give out. This is a secret to finding happiness in life.
12. **Balance:** Seek balance in all areas of life, considering family, relationships, career, business, health, leisure, and spirituality.
13. **Gratitude:** Be thankful for the gift and opportunity of life. If you have a spiritual belief, pursue the core tenets of the teachings and make them part of who you are.
14. **Find Your Excitement:** Find things in life that excite you, and act on them. Embracing things which excite and motivate you will move you more quickly into success and happiness, and attract the things you need to get there. This attitude will support and integrate with all the above tenets.

Personal Development Planning

Personal development planning is the process of creating an action plan based on awareness, values, reflection, goal-setting and planning for personal development within the context of a career, education, relationship or for self-improvement.

The PDP (personal development plan), also called an IDP (individual development plan) or PEP (personal enterprise plan), usually includes a statement of one's aspirations, strengths or competencies, education and training, and stages or steps to indicate how the plan is to be realized. Personal development plans may also include a statement of one's career and lifestyle priorities, career positioning, analysis of opportunities and risks, and alternative plans and Curriculum Vitae.

In higher education, personal development plans typically include a portfolio containing evidence of the skills gathered over a particular timeframe. It is presumed in education that undertaking PDP will assist in creating self-directed independent learners who are more likely to progress to higher levels of academic attainment. It is also used in Human resource management.

Personal developments plans are often a requirement for employee CVs. Employees who are participating in business training are often asked to complete a personal development plan.

A five year personal development plan can often be developed by an individual to organize personal goals and make them achievable within a certain time period.

Focus on what is truly important by using the Quick and Easy Personal Development Planning

Good personal development planning will help you achieve your potential as it will help you identify skills gaps and improvement areas. Then you'll be able to get that job, go for that promotion, build your business further, or whatever you want.

This guide will give you a simple yet powerful method to grow and so enable you to achieve more, faster, and more effectively.

Personal development planning can be daunting, but with the right approach, it can be highly satisfying, and even fun! Use this guide with built-in template to plan your personal development properly.

Personal development planning can be daunting, but with the right approach, it can be highly satisfying, and even fun! Use this guide with built-in template to plan your personal development properly.

"If your personal resources don`t match your goals and ambitions, it`s time for some personal development"

Some detail on what the Personal Development Planning covers.

- Personal Development Planning Contents
- Making dreams into reality;
- The 10,000 foot view;
- Phase 1 - What's Your Focus?
- Phase 2 - Using the Project YOU planning template;
- Phase 3 - Review the Project YOU Development Plan;
- Phase 4 - Implement the Development Plan;
- Phase 5 - Review progress;

Personal Development Plan Template

"Get out of that rut take control, focus on the important things".

Here's some detail on what the Goal setting covers.

- The Goal Setting Contents
- Why set personal goals?
- Benefits of personal goal setting;
- Goals and your personal values;
- Writing your goals;
- Smartness in goals;
- Example Goal Sheet;
- Background Thinking and Processes Used to Create the Goal Sheet;
- Results review and link to self assessment.
- Personal Development for Smart People

The purpose of this website is to help you grow as a conscious human being. This includes guiding you to discover and accept your life purpose; inspiring you to feel more motivated, energized, and passionate; helping you shed disempowering relationships and build a network of loving support; teaching you how to achieve stable financial abundance doing what you love; and encouraging you to make a genuine contribution to humanity -- so you can finally experience the kind of life that deep down, you always knew you were meant to live.

You aren't here to struggle and suffer. You're here to express and share your creative gifts, to give and receive love, and to be happy. It will take time, but this site can certainly help you get there, and the vast majority of resources here are free.

This site will help you make conscious decisions in your personal development journey and courageously follow through. This means having the maturity to take 100% responsibility for your health, your career, your finances, your relationships, your emotions, your habits, and your spiritual beliefs. It requires taking a deep look at yourself, consciously deciding what kind of person you truly are on the inside, and then getting your external reality to be congruent with your inner being. The goal is to help you achieve outstanding effectiveness while maintaining internal balance, where your thoughts, feelings, actions, and skills are all working together to create the life you truly desire.

Personal development is hard work. It takes time, consistency, and patience. If you're only here looking for shallow quick fixes and you aren't interested in real, lasting change, this isn't the right place for you. This site is designed for people who are serious about personal growth and willing to commit to it.

The main themes of this topic are:

- **Truth:** Accept reality and rid your life of falsehood and denial
- **Love:** Improve your ability to connect with yourself and others
- **Power:** Build your motivation and discipline to create the life you desire
- **Oneness:** Stop fighting against resistance, and make the world your ally
- **Authority:** Take command of your life, and learn to make clear decisions

- **Courage:** Summon the inner strength to take action in spite of fear
- **Intelligence:** Live authentically, and express your genuine creativity

The areas of your life this topic will help you improve include:

- **Habits:** Create a daily routine that gives you a sense of flow
- **Career:** Build a career you're truly passionate about
- **Money:** Achieve financial abundance without compromising your integrity
- **Health:** Adopt health habits that empower your physical body
- **Relationships:** Enjoy loving relationships that help you grow
- **Spirituality:** Raise your awareness and live more consciously

Many people report that it helps them become more focused, motivated, and confident.

- The Courage to Live Consciously;
- Personal Development Articles;
- Personal Development Blog;
- Most Popular;
- Personal Development Podcast;
- Personal Development Newsletter

Twelve Personal Development Tips for Life Success

Personal development is well-known idea that many people aim to use and accomplish something from. Every time you decide to take that new course or new training you are developing some aspect of you. That aspect might be to acquire a new skill or improve an existing skill. It might be to become more knowledgeable. It can even be to change some fundamental part of you such as a behavior or block you might have that limits or stops your route to success.

Anytime you aim to change something about your life and how you operate within your world, this could be deemed as personal development. Even unwanted and seemingly negative experiences can form part of our understanding and experience of the world. Through that negative or upsetting event you will inevitably change and therefore develop. So, our self development can happen through positive and negative experiences. It can happen through self-determined as well as what might be termed fate-determined events.

Here are some tips for the pro-active people who like to take charge of their lives and who choose to develop their potential. Feel free to add more of your own tips or break down each tip into smaller steps to suit your needs and circumstances.

1. Writing a plan, as is so often acknowledged, increases the odds of you actually carrying out that plan. So don't keep it in your head.
2. Now acknowledge your past successes. No matter how long ago those successes might be, it is a good idea to be reminded of what success is and how rewarding it feels as an experience. Write down the successes that have meant the most to you.
3. Now look at all the areas of your life which include as they say, work, rest and play. So look at your occupation, your home life, and your recreational pursuits and write them all down. These must be what you are currently doing.
4. Now rate how happy or successful you are or how much you enjoy those areas you wrote down for no. 3)
5. If there are any areas from no. 3) that stick out as being energy-sapping, unhappy things or activities that you endure, or aspects of what you do that you could improve, than these could indicate a focus for your energy regarding your personal development.

6. If you are happy with everything in no. 3) or for whatever reason choose not to change anything right now, than that is okay. It at least means you have reviewed your current circumstances.
7. If tip no. 6) holds true for you, then you may be considering adding to what you already do and how you utilize your time, energy and resources. If you know what it is you want to work on, great. If you don't know, do a little day-dreaming and just let your mind flit between ideas and desires until you settle on some change that excites you. A change for that you are passionate about.
8. Whatever area of your life you have chosen to change or add to and therefore to develop, it is vital that you feel motivated to change. So this change must be chosen by you and not because someone else thinks it is the right thing for you to do. You must "own" your goal for change if you are to see it through and enjoy success.
9. Now write down on your plan, in vague terms initially, what your personal development desire is.
10. Now you need to describe in more detail your practical ideas for turning your stated desire into your reality. What steps do you need to take to make your desire become a reality?
11. Take the steps you have written down and write target dates beside each one. This will keep you focused on getting the steps completed and help keep you motivated as well as providing evidence that you are progressing.
12. You may need to review and refine what you do as you go along. Flexibility and a willingness to adapt to unforeseen events and changing conditions will enable you to succeed.

You also need to know that you may have parts of your personality, past influences and habitual behaviors and deep-rooted attitudes that can block and sabotage your progress during any of the twelve steps.

9

Self-Discipline

What is self-discipline?

- Self-discipline is self-control and self-restraint.
- Self-discipline is self-reliance and independence.

Students who exercise self-discipline:

- Complete their assignments
- Stay on task
- Wait to be called on
- Work toward personal and community goals
- Try again and again
- Ignore peer pressure
- Choose productive rather than destructive activities
- Control their tempers

Eight ways to strengthen your self-discipline

1. Decide that you really want to be someone who's self-disciplined. Your desire will motivate you to make good choices.
2. Make a personal commitment to develop and strengthen these traits. Write down specific things to do.
3. Learn the rules that determine what you can and cannot do.
4. Be accountable. Accept responsibility for your own behavior. Don't blame others for your actions and decisions.

5. Practice. Self-discipline is something you can teach yourself. For example, set aside time to read more or to clean up.
6. Do activities that enhance your self-discipline like yoga, walking, rock-climbing, practicing a musical instrument.
7. Eliminate harmful habits. For example, if you spend several hours each week watching violent videos or TV programs, make a conscious decision to spend your time in healthier, more productive ways.
8. Start a self-discipline group to plan and carry out activities.

Activities

- ❍ Brainstorm a list of problems that might result from a lack of self-discipline: personal appearance, physical-mental-emotional health, school success, life success, friendships, job performance, talents, and participation in family-clubs-community-faith, marriage, and parenting.
- ❍ Do silly exercises to strengthen your self-discipline like not scratching.
- ❍ Role-play how you might talk with a younger brother or sister who is demonstrating a lack of self-discipline.
- ❍ Write or tell a chain story about a make-believe prince or princess who has no self-discipline.
- ❍ Interview scientists, engineers, doctors, and other people to learn the role of self-discipline in training for their careers and in working.
- ❍ Investigate different animal species to learn if they use discipline. Is misbehavior punished in young or in a group situation?
- ❍ Examine nature to find examples of discipline and order such as patterns.
- ❍ Explore tessellations (repeating geometric patterns).
- ❍ Research discipline in history.
- ❍ Draw cartoons showing examples of self-discipline vs. no discipline.

- Write new lyrics to a popular song that encourage self-discipline, self-restraint, and self-reliance.
- Explore musical dynamics and control.
- Learn about self-discipline in sports.
- Play a "stop-wait-go" game thinking of different situations.

What is Self-Discipline-Definitions

Self-discipline appears in various forms, such as perseverance, restraint, endurance, thinking before acting, finishing what you start doing, and as the ability to carry out one's decisions and plans, in spite of inconvenience, hardships or obstacles.

Self-discipline also means self-control, the ability to avoid unhealthy excess of anything that could lead to negative consequences.

One of the main characteristics of self-discipline is the ability to forgo instant and immediate gratification and pleasure, in favor of some greater gain or more satisfying results, even if this requires effort and time.

The term self-discipline often causes some discomfort and resistance, due to the erroneous notion that it is something unpleasant, difficult to attain, and which requires a lot of effort and sacrifice. Actually, exercising and attaining self discipline can be fun, does not require strenuous efforts, and the benefits are great.

True self-discipline is not a punitive or restrictive lifestyle as some people think, and it has nothing to do with being narrow minded or living like a fakir. It is the expression of inner strength and staying power, vital for dealing with the affairs of daily life and for the achieving of goals.

Self discipline, together with willpower, can help you overcome laziness, procrastination and indecisiveness. Both enable you to take action and persevere with it, even if the action is unpleasant and requires effort.

With Self-discipline you can exercise moderation in what you do, become more patient, tolerant, understanding and considerate. In addition, it helps you withstand external pressure and influence.

A self disciplined person is more punctual, and invests more time and effort in what he or she does.

A self disciplined person is more likely to take control of his or her life, set goals, and take concrete steps to achieve them.

Self discipline is well portrayed in the story about the rabbit and the turtle, who conducted a race. The rabbit was very he was faster, so he allowed himself to take a nap in the middle of the race. At same time, the turtle plodded along, but with willpower and self discipline him and eventually managed to arrive first to the finish line. Like the turtle, with self discipline you finish what you start.

Strengthening willpower and self discipline required for achieving any kind of success

"Self-discipline begins with the mastery of your thoughts. If you don't control what you think, you can't control what you do. Simply, self-discipline enables you to think first and act afterward." Napolean Hill

"Discipline really means our ability to get ourselves to do things when we don't want."—Arden Mahlberg

"Self-discipline is a form of freedom. Freedom from laziness and lethargy, freedom from the expectations and demands of others, freedom from weakness and fear-and doubt, self-discipline allows a pitcher to feel his individuality, his inner strength, his talent. He is master of, rather than a slave to, his thoughts and emotions."

The Five Pillars of Self-Discipline

The five pillars of self-discipline are: Acceptance, Willpower, Hard Work, Industry, and Persistence. If you take the first letter of each word, you get the acronym "A WHIP" — a convenient way to remember them, since many people associate self-discipline with whipping themselves into shape.

Each day of the series, I'll explore one of these pillars, explaining why it's important and how to develop it. But first a general overview....

What Is Self-Discipline?

Self-discipline is the ability to get yourself to take action regardless of your emotional state.

Imagine what you could accomplish if you could simply get yourself to follow through on your best intentions no matter what. Picture yourself saying to your body, "You're overweight. Lose 20 pounds." Without self-discipline that intention won't become manifest. But with sufficient self-discipline, it's a done deal. The pinnacle of self-discipline is when you reach the point that when you make a conscious decision; it's virtually guaranteed you'll follow through on it.

Self-discipline is one of many personal development tools available to you. Of course it is not a panacea. Nevertheless, the problems which self-discipline can solve are important, and while there are other ways to solve these problems, self-discipline absolutely shreds them. Self-discipline can empower you to overcome any addiction or lose any amount of weight. It can wipe out procrastination, disorder, and ignorance. Within the domain of problems it can solve, self-discipline is simply unmatched. Moreover, it becomes a powerful teammate when combined with other tools like passion, goal-setting, and planning.

Building Self-Discipline: -My philosophy of how to build self-discipline is best explained by an analogy. Self-discipline is like a muscle. The more you train it, the stronger you become. The less you train it, the weaker you become.

Just as everyone has different muscular strength, we all possess different levels of self-discipline. Everyone has some — if you can hold your breath a few seconds, you have some self-discipline. But not everyone has developed their discipline to the same degree.

Just as it takes muscle to build muscle, it takes self-discipline to build self-discipline.

The way to build self-discipline is analogous to using progressive weight training to build muscle. This means lifting weights that are close to your limit. Note that when you weight train, you lift weights that are within your ability to lift. You push your muscles until they fail, and then you rest.

Similarly, the basic method to build self-discipline is to tackle challenges that you can successfully accomplish but which are near your limit. This doesn't mean trying something and failing at it every day, nor does it mean staying within your comfort zone. You will gain no strength trying to lift a weight that you cannot budge, nor will you gain strength lifting weights that are too light for you. You must start with weights/ challenges that are within your current ability to lift but which are near your limit.

Progressive training means that once you succeed, you increase the challenge. If you keep working out with the same weights, you won't get any stronger. Similarly, if you fail to challenge yourself in life, you won't gain any more self-discipline.

Just as most people have very weak muscles compared to how strong they could become with training, most people are very weak in their level of self-discipline.

It's a mistake to try to push yourself too hard when trying to build self-discipline. If you try to transform your entire life overnight by setting dozens of new goals for yourself and expecting yourself to follow through consistently starting the very next day, you're almost certain to fail. This is like a person going to the gym for the first time ever and packing 300 pounds on the bench press. You will only look silly.

If you can only lift 10 lbs, you can only lift 10 lbs. There's no shame in starting where you are. I recall when I began working with a personal trainer several years ago, on my first attempt at doing a barbell shoulder press; I could only lift a 7-lb bar with no weight on it. My shoulders were very weak because I'd never trained them. But within a few months I was up to 60 lbs.

Similarly, if you're much undisciplined right now, you can still use what little discipline you have to build more. The more disciplined you become, the easier life gets. Challenges that was once impossible for you will eventually seem like child's play. As you get stronger, the same weights will seem lighter and lighter.

Don't compare yourself to other people. It won't help. You'll only find what you expect to find. If you think you're weak, everyone else will seem stronger. If you think you're strong, everyone else will seem weaker. There's no point in doing this. Simply look at where you are now, and aim to get better as you go forward.

Let's consider an example.

Suppose you want to develop the ability to do 8 solid hours of work each day, since you know it will make a real difference in your career. I was listening to an audio program this morning that quoted a study saying the average office worker spends 37% of their time in idle socializing, not to mention other vices that chew up more than 50% of work time with unproductive non-work. So there's plenty of room for improvement.

Perhaps you try to work a solid 8-hour day without succumbing to distractions, and you can only do it once. The next day you fail utterly. That's OK. You did one rep of 8 hours. Two is too much for you. So cut back a bit. What duration would allow you to successfully do 5 reps? Could you work with concentration for one hour a day, five days in a row? If you can't do that, cut back to 30 minutes or whatever you can do. If you succeed, then increase the challenge.

Once you've mastered a week at one level, take it up a notch the next week. And continue with this progressive training until you've reached your goal.

While analogies like this are never perfect, I've gotten a lot of mileage out of this one. By raising the bar just a little each week, you stay within your capabilities and grow stronger over time. But when doing weight training, the actual work you do doesn't mean anything. There's no intrinsic benefit in lifting a weight up and down — the benefit comes from the muscle growth. However, when building self-discipline, you also get the benefit of the work you've done along the way, so that's even better. It's great when your training produces something of value AND makes you stronger.

Throughout this week we'll dive more deeply into the five pillars of self-discipline. If you have any questions on the subject of self-discipline (either specific or general) that you'd like to see addressed, feel free to post them as comments, and I'll do my best to incorporate them along the way.

Personality Development and Positive thinking

While you are on your path to develop your personality, do it on a positive note and you will feel a world of a difference in the new you. Here are some techniques that can help you develop your personality and experience the effect of positivity in your life. Before you read on or do anything else, are happy with your present self and embrace yourself completely. Remember you are only one of a kind.

Commit yourself

For anything you plan to do, whether it is a job, writing a book or learning a new language you have to commit yourself to it. Commitment comes in the form of the doing the thing you want t do for a long period of time or very often. So if you want to be more positive you have to commit yourself to positivity. You have to start thinking positively and banish negative thoughts. Begin with your mind. Cleanse it. Since you are reading this commit yourself to personality development since personality does not star at the end of this article, any other course or book but it is an on-going process. So keep working on yourself. Remember, 'the road to success is always under construction'.

Focus

Focus on the activity you are doing. True yoga is being done with the activity you are into at that moment. So if you are wearing clothes pay complete attention. If you are eating, do only that. Love your food completely. If you are washing vessels, be lost in the scrubbing. If you are talking to someone be present, don't talk absent-mindedly. While you exercise, focus on the steps. Be one with your positive thoughts.

Do not give up

Never ever give up. Things can go wrong, mistakes happen and no one is flawless. So keep going. If you fall, rise and try again. Don't let failures be a deterrent to your success. If you stop trying you put an end to your goal. But if you keep trying there are chances you will succeed. Do it with faith and believe in you. More than intelligence your persistence will take you places.

Keep your ears and eyes open

Always keep your eyes and ears open. Listen to what people say. Observe whatever is around you. Never give anything or anyone invisible treatment. Being alert helps you learn and keeps you informed. So if you are having a conversation with someone the next you know that you are in touch with your surroundings and can discuss anything under the sun. If you do not know something, say so. Asking questions that shows you have interest in the discussion and are listening.

Love yourself

If you are looking forward to develop your personality, the first thing you have to do is love yourself. Accept yourself. Accept everything that's a part of you - your skin, your body, your ideas and everything else that makes you, you. Take care of yourself. Feed yourself and have a well- nourished body. Groom yourself and be well-dressed. Don't forget to wear a smile. Be calm and relaxed. Mediate to calm down. Keep anger under check. Exercise regularly and engage in prayer. Accomplish your work and ear compliments.

Personality development happens first from inside and then outside. So start treating your mind and body with special care. Develop positive thinking and turn into a new shining personality.

The Power of Positive Thinking

Ever wondered why Mahatma Gandhi said, 'Your thoughts are as good as your deeds'? Maulana Abdul Kalam former President of India writes in his autobiography, 'Wings of Fire', that thoughts have energy. Whatever we think has the power to become true. Our thoughts are energy - Energy which goes out into the universe and conspires to turn thoughts real. No wonder why prayers are answered. No wonder why ideas see the light of day in the form of real moving mechanisms, structures and monuments. Even the Taj Mahal was just a wish before it turned a wonder of the world.

Attracting achievements

Positive thinking is a powerful tool to attract all that you want towards what you want to achieve in your life. It all begins with a thought. When you were young you may have thought that you want to become a teacher. Your thoughts grew and as you kept growing you positively believed that you are going to be a teacher one day. You attracted things towards yourself that made it possible. There are many times people do things that may be beyond their capacity or potential. But they do accomplish task because they lead themselves with a belief of being able to do things and think positively. Imagine if Mahatma Gandhi sat sulking in a corner saying we could never achieve freedom. Or if Sherpa Tenzing thought that Mt. Everest is too high for him. Behind every task that seems impossible is a positive belief that attracts things that can make the impossible possible.

Living and no just surviving

The power of positive thinking is your power to live and not just survive. Once you realize the power of positive thing you can harness the life that you desire from it. Positive thinking keeps you in a mind frame that is confident, courageous and makes you feel happy even in adversity. It is a service that comes to you for free. Positive thinking can turn your life into a happy, prosperous and peaceful one if that is what you desire. There is nobody other than you who can think positive for yourself. It is one basic step that can help all facets of your life and make you a better human. Lead yourself to think positively than just live each day complaining about why life is unfair to you.

Allow positive thinking in your life

You are the leader of your life. You have to control your thoughts and allow positivity in your life. Famous author Robin Sharma advises to treat your mind like a garden. The negative thoughts are the weeds that are running the garden and even making you turn bitter. To let positive come in your life first pull out these negative weeds and make room for positivity. Let all your negative thoughts go. Just let go. Replace each negative thought with a positive one. As soon as you get a negative, stop yourself and let go. Now think a positive one. Slowly you will form a habit of thinking positively. Before you know you will be far more positive, happy and successful.

Let the power of positive thinking be yours

While you start seeing changes in your life do not give up thinking positive. Keep reading about positive thinking. Share your ideas with others. It will only make your belief stronger. Helping other stay positive will make you more positive and grow your power to remain positive and strengthen your belief system. You will be able to deal with tough situations easily. If at times you feel down in the dumps come back to this page and refresh everything that you learnt about the power of positive thinking. Let the power of positive thinking be yours forever.

The Power of Positive Thoughts

Each of us is empowered by something that lies within us to realize our true potential and purpose. That power is nothing else but the power of positive thoughts. Every thought we think influences our lives. These thoughts could be our own or thoughts given to us by someone else. It is entirely in our control how we think to steer ourselves on the path we want. Every thought has the capacity to come true. Thoughts have energy and this has been proven scientifically too. The first step towards have a perfect life is to first cultivate perfect thoughts.

Going beyond negative thoughts

Cultivating positive thoughts is not easy when your mind is filled with negative suggestions and thoughts. These negative thoughts can cloud your thinking and send negative messages to your body leading to bitterness. Negative thoughts are like weeds that ruin the garden of your mind. So it is very important that you move beyond negative thoughts first.

The easiest way to do it is to stop them the moment you start off. Simply stop it. Let go of all the negativity. Replace negative thoughts with positive ones. Once you start doing this you will form a habit of thinking positively.

Create values

Live by your values. Think positive thoughts. Extend them with the right actions and soon you will form your own positive values. Acting in a positive manner in various situations helps you build positivity. This will first begin with a positive thought that you translate into action and words. E.g. If you are unhappy with your spouse don't think negatively about his or her behaviour, choose to think a better thought. Think on his/her behalf. Choose to react positively instead of negatively. Choose to treat the situation with a positive approach and you will surely get positive results. If you are suffering from a health problem you could choose to feel better or you could simply feel like a victim and worsen your condition. Just remember every time you get react negatively, get negative thoughts or judge someone stop right there and alter your thought. Soon it will be a value you will treasure all your life and the one that give you the life you desire.

Influencing your life

Start influencing your life with the power of positive thoughts. After hearing the story of Aladdin and the Magic lamp most of us wish that we had such a lamp with us that would give us our heart's every desire. Well the magic lamp is within us. We have just forgotten to use it enough to our benefit. That little lamp is the lamp of positive thinking.

Positive thoughts will make this lamp shine brighter and fulfill your commands and soon you will experience magic in your life. Our lives are influenced by so many people and situations; we can surely take one step towards thinking for ourselves. You can shape your life by the simple means of choosing your thoughts and exercising the power of positive thinking.

The Benefits of Positive Thinking

Positive thinking is not just a technique to keep you going in times of adversity. Positive thinking in your daily life can have profound benefits for you and how your life shapes up. Once you start embracing positive thinking you may find so many things the way you wanted them to be or may be even better.

A will to heal with positive thinking

You may have seen people who heal very fast from a health problem and some who may not show any progress. The ones that step out of their problems and take to life are the ones who have a will to recover. Recovery is always faster in people who positively want to heal. You may be positive that an injection will not hurt you and surely you will find it was just a pinch. Thinking negatively about pain makes the pain worse. Similarly for women in labour who are calm and concentrating with a positive belief that they will deliver safely with minimum pain are the ones who enjoy a good delivery. Well, sometimes it also happens that women told by medical practitioners that they can't have a baby, get pregnant defying medical reasons because of the mother's positive will. It is only what you think that makes up your reality.

A better body with positive thinking

Positive thinking about your body and loving it the way it is allows your body to be good in return. If you positively think what you want to change about your body it will surely happen. Your body works hard for you, so appreciate it and avoid negatively blaming your body for being fat, overweight or being in disgust with your thighs or wrinkles. Stop the obsession and start loving yourself and you will experience health benefits that you desire. You will stop over-eating and you will work towards a healthy body with optimum weight. Think about your body as a temple where your soul lives. Take care and think positively about the changes you want to make. Love your beautiful body for it has served you all these years.

A better career with positive thinking

While you are working hard you believe you will never reach the top because of your colleague who is closer to your boss. Or your boss who takes away all the credit for the work had done by you. You hate your office administration, you blame the management and you simply would think of leaving the whole bit if you got a better job where you would get promoted faster and earn more. Well, have you ever thought of how you are contributing to the success of the organisation? Have you ever checked your faults? Have you ever tried to show compassion to your colleague instead of competing with him? Have you owned responsibility for your mistakes or have you ever thought of talking to your boss about your growth? Have you ever been positive about the place that gives you a pay cheque every month? Instead of playing the blame game, be positive about your work. Think of things you could improve. Be positive when you ask for an appraisal. Check what extra can you do beyond your job. Start thinking positively and you will love your job and even get better returns.

A better parent with positive thinking

Instead of blaming your children for being naughty, be positive that they will grow up to be great men and women. Allow yourself to trust them. Give them a chance. Remove negative fears about their education from your mind and be with them. Share your positive thinking with them. Read stories of positive thinking together. Watch inspirational CD's. Build a rapport.

Achieve excellence with positive thinking

Whatever you want to do with your life, remember you have the potential and the capability. You have to silence your fears, make an intention to achieve your goals and positively fuel your thoughts to light a reality that you desire. Remember, 'if your heart believes and your brain conceives it will happen.'

Positive Thinking Techniques

While you have read and heard so much about positive thinking and you want to embrace the power of positive thinking you do not know how you can do it. Well the techniques of positive thinking are simple to understand and follow. However they require persistence from you. It will be of no use if you give up practicing the techniques of positive thinking. These techniques are a step towards entering a prosperous life and fulfilling the desires of your heart. You can go through these techniques and start off your own positive thinking program.

Write it down

Write down your goals or affirmative thoughts on paper. Writing your goals on paper gives them life. Keep going back to it regularly and work towards fulfilling them. Feel good about yourself and write down what you want to be or how you want to see yourself. Instead of writing, 'I want to be happy', write 'I am happy', 'I am beautiful', 'I am healthy' or 'I am committed'. Write down all that you want to be in a way that you already possess it. These will influence positive thoughts and drive you towards achieving what you desire.

Be Mindful

Be mindful in all the activities you do or at least try being mindful in some activities to begin with. There are times when you are reading something and remember nothing once you are done. This is because you have been reading mindlessly. Mindfulness requires concentration. When you read something read it will all the concentration. Be one with the activity. When you are eating something feel the taste, the texture and how the food is going into your system and taking you from the stage of being hungry to being satiated.

Observe your surroundings completely, look for minute details. Feel your bed when you sleep. Feel how you sink into restful sleep, touch the pillow and feel the sheets. Similarly feel your thoughts. See how you feel different, happy and more peaceful with positive thoughts. Examine how it takes you from the feeling of bitterness to a feeling of rising above your problems and feeling good about yourself.

Talk to yourself

Talk to yourself about negative thoughts. Scream, shout and remove it out. Don't carry burden that you do not require. Look in the mirror and embrace all that you think are physical flaws in your body. They are you. Think of a positive change but yourself while doing so. Forgive yourself and be kind to yourself. Love yourself before you love anyone else. Reward yourself with relaxation treatments and spend time in nature. Accept that you are confident, lovable and valuable. Let yourself know it.

Meditate

The most important thing to do to attract positive thinking in your life is to meditate. It might seem difficult first but it is possible. Find a quiet place to sit and sit there to meditate. Ensure you are wearing comfortable clothing and have no clutter around you. You can concentrate on your breath and even make use of chants like 'om' or the sound to meditate better. Meditation clears and calms your mind.

Let go

This is the most important technique of attracting positivity in your life. Simply let go. Whether someone has deceived you, you have faced financial loss, your boss screamed at you, you had a break off or things did not happen the way you had planned them, simply let go. Letting go frees your mind and soul of unnecessary burden that you carry along adding to miseries. So when there is something bothering you just let it go. Think of a positive thought and move on.

Positive Thinking Exercises

Thinking positive may seem very difficult after all those years of negative judgments and statements. Something that has been in our psyche for so long is not easy to give up in a day. However, there are a few positive thinking exercises that can be your guide in building your very positive thoughts influencing a life that you desire. Here are some of them to let you make a start and keep going with positivity.

Breathing exercises

Begin your morning or end your day with breathing exercises like the pranayam. The pranayam is a yogic way of breathing in which you place your two fingers in the middle of your eyebrow, thumb on the right side of the nostril and ring finger on the left side of the nostril. Begin with the left side. Lift the finger inhale. Place the finger back and count till five and exhale from the right nostril. Now breather in from right, wait and breathe out from the left. One set of pranayam is complete. Practice 10- 20 sets and relax. This way of breathing relaxes and even re-energises your mind. Working with a calm mind helps you be more positive.

Physical Exercise

Do some physical exercise. Move about the office. Take a break and go over to your colleague's desk and have a short chat. You may share a joke that may even make you laugh. Practice yoga or any other form of exercise like Kick boxing or dancing. Exercising releases hormones called endorphins that make you feel happy. You are always more positive and feel lucky when you are happy So always walk a bit or do a bit of a jog to keep the adrenaline pumping. Besides, it will also keep you in great shape.

Get some sunshine

Move out in the sun. Feel the heat. Sweat a bit and feel you alive. Let some sunshine in your room. A bright environment always keeps you cheerful than the one that is dark and gloomy. How do you feel when you simply draw the curtains and the rays of the sun simply light the room up? Doesn't it perk you up?

Practice Gratitude

Always, always practice gratitude. While you always crib about why you were neither born in a millionaire family nor have a millionaire for a spouse. Or why you don't own a flashy sedan? Rather think what if you were born in Afghanistan or face what the people of Iraq face each day. Or why you were not in the twin tower on business on the day of the fateful incident or why were you not in Miami when an earthquake occurred.

There are thousands of people who go through worse, who are not sure when they would get their next meal. There are people who are working on lowly wages, people who working in inhuman conditions so that your surroundings remain clean and there are the soldiers who stay put on the borders for our safety while their families pine for them.

Appreciate the meal you have in front of you, appreciate your friends, appreciate that you have a cosy bed to sleep in, show gratitude for a peaceful day and love yourself for who you are. You are far better looking, well-fed and comfortable than many other people on this planet. Think about it and you will surely switch to a positive frame of mind.

Persistence

Never give up. Follow your goals with persistence. Have a 'come what may', attitude and keep going till you reach where you want to. If you were slip up on your exercises of positive thinking come back to this paragraph and start persisting once again, you will catch up fast.

Positive Attitudes

A positive attitude can take you far ahead than you can even think. A positive attitude can help you overcome crisis, stay your ground and a strength to go behind your dreams. A positive attitude is like a wing that can help you fly where you want. So if you always wanted to be like a bird now you know you just have to sprout the wings of a positive attitude and sky is the limit for you.

Shape your personality

Who says that looks do not matter? A clean and neat looking person is always preferred than a person who is ill-dressed and not groomed enough. A polite speaker is always better than someone who guffaws loudly. So it is important to shape your personality. A nice personality with a positive attitude is what everyone seeks. So learn to talk well. Be polite and listen to others. Have your questions but give attention to answers given by others. Avoid ridiculing. Give respect to everyone. If you are not a good speaker work on it and you will learn soon.

In a team

Never get bogged down by working in a team. Respect your team mates and make room for everyone, even for yourself. Give respect and address others mistakes with politeness. Do not compete within. Think of the good of the team. Cultivate feelings of mutual understanding by sharing things and taking responsibility. Avoid being a one man show person. If there is a problem discuss it openly with the team but avoid anger. Look at sorting out the issue and not proving yourself right.

At home

Work with a positive attitude at home too. If your child breaks something do not scream. Try mending it together or explain to the child that he or she must be careful because it costs money and it's hard to earn. Be cool if your husband or wife is angry. Talk to them when they finish talking or when the storm has subsided. Seek an apology if it is your mistake. Forgive them if they are wrong. Do not be revengeful. For men and women who complain of ill treating in-laws can try stepping into their in-laws shoes. Respect them but do not cover their faults. Avoid anger, fights and complaining. Relax and think of a positive way out of your problems.

With yourself

Do not be tough on yourself. Forgive yourself for making mistakes. Love yourself for what you are. Avoid comparing yourself with others. Show gratitude for what you have. Take each day at a time. Focus on the current activity and keep negative thoughts out. Stop dwelling on the past and live today. Love your body and nourish it. Eat well and live healthy. A daily dose of meditation can keep you away from medications. All these things help you build a positive attitude and help you stay happy.

Staying afloat in adversity

When in adversity do not panic. Stay afloat and look for solutions. A calm mind is better equipped to take decisions than a mind in distress. So if adversity strikes you and things aren't working well, believe in yourself and move on. Shoo away your fears and have faith. Once you are above the problem you will experience the magic of a positive attitude.

Motivation and positive thinking

For everything in life you need motivation. For working hard you need the motivation of a good remuneration. Peers engaging in an activity can serve as a great motivation to do the same thing. Peers are always a great motivating factor. Motivation is a stimulus to do things. So to remain positive and continue living with positive thinking you have to keep up the motivation to do so. There are a number of ways that can motivate you to remain on the positive side.

Watch and Learn

Taught to children, the watch and learn method is a great way to motivate yourself and inspire you to lead a great life with positive thinking. Simply watch what great men and women do. Read about great lives. Pick an autobiography of Richard Branson, Mahatma Gandhi, President Barack Obama, and Kishore Biyani, President Abdul Kalam or any other personality that you admire and find how they led their lives. What motivated them to reach their ultimate goals?

You can even watch what your successful colleagues do. There is always a set of values and standards that a successful person follows. There are motivating factors that keep them going. Find them and then find what could be a great motivating factor for you. Once you search you will surely find.

Keep in touch with yourself

Never lose touch with yourself. Always keep introspecting. Find what do you like doing, what do you enjoy the most. It will help you lead yourself towards success. These likings could be great motivators to keep you moving ahead. E.g. If you enjoy dancing more than running on a treadmill then join a dance class that you enjoy and losing weight will be a bonus. Besides that you won't even have to drag yourself to an activity you do not enjoy. You will find yourself happily attending the classes and looking forward to sessions. Once you lose enough weight, your new body will motivate you to continue doing what you were doing.

Get inspired by Nature

Nature can be a great motivator. Try visiting gardens or walk through a place where there are lots of trees. You will like it. You can drive up the hills and enjoy the nature. Go for a trek. See how the mountains stand tall uninhibited buy any obstacles. Look at the sea, how the far and wide it spreads itself. Look at small things like a honey bee that works hard to make honey. Look at the silk worm it works hard so that you can wear a silk dress. How about the rains that quench the thirst of the parched earth? A closer look at nature can serve you hundreds of examples that can motivate you for life.

Men and women in History

Look and observe great men and women in history. Great men and women who have become shining examples they will inspire generations to come. People who did things that none of us can match but aspire to come closer to. They are Gautam Buddha, RamasesII, Jesus Christ, Emperor Akbar, Cleopatra, Mother Teresa, Florence Nightingale, Anne Frank, Jahangir Tata and Abraham Lincoln.

All of them have a different background and did different things for the betterment of humanity. They are the epitome of passion and dedication towards the things they believed in and desired to achieve. A little about information about them can serve as great motivation for you to do something that is for the greater good.

Compete with yourself

Stop competing with anyone else. Not just in your office but stop competing with relatives and friends. Just compete with yourself. Avoid developing negative materialistic obsessions thinking they are motivators. E.g. If your neighbour has bought a Porsche, you need not take that as a motivator to push yourself to buy the next best car only to show that you too have it in you. If you have the money great, but why lose hard earned money on trifles. This will only add to your agony. Rather feel good for your neighbour, feel happy that something good happened for him.

You do not have to prove yourself to anyone. Just be true to yourself and the world will know. People can always see through. Make out the difference between a motivator and a negative obsession. Always step back if you can't decide. You will know. Love life, get inspired and reach out to your destiny, with positive thinking and loads of inspiring motivation.

Negative thoughts and sadness

Do you feel like laughing around a person who is sad and crying? Do you enjoy the company of people who are constantly complaining or judging others? It's like the yawning phenomena, when one yawns the others get infected too. Very similar to a smile or laughter! None likes to be around a grouch. Therefore it is good to get over things that haven't worked according to what you planned and smile in the face of adversity. Getting stuck to negative thoughts only increases the problem. Every thought has the potential to come true. Would you want to trade this fact by having negative thoughts rule your head and life?

Blow away the blues

Watch a funny film, watch cartoons, play with kids or read a book. It will pep sooner than you think. Reading motivational books can be of real help. Go meet an old friend and share the problem. It will lighten your heart and you may even find that the problem has a funny side to it. Problems are like lions if you don't learn to ride them they will eat you up. Negative thoughts are like parasites they feed on you. Its best to clean them up and drive them away. So if you are feeling sad and having strong negative thoughts it's not going to blow away the blues. Just let them go and are quiet for some time. Meditate and be still, it will help.

How negativity eats you

Negativity eats you bit by bit. You have a few negative thoughts and if you keep thinking on the same tangent they snow ball into very strong negative thoughts of jealousy, hatred and cruelty. Your mind will go on with the chatter, making you feel worse and sapping you off your energy. You will lose focus on the things you have to do and feel tired very fast. Uproot negative thoughts and it's as easy as adding a full stop to a sentence. Yes! It is that easy, you got to try it. Just ask yourself to stop. Say it aloud if you want. Replace the thought with a positive one. Give up your pre-designed fears. Breathe.

Don't feed negativity

Never feed negativity. If you are already having negative thoughts about something or someone, avoid sharing it with others. They might influence you further. Just write your thoughts on paper if you want them out of your system and tear the page or burn it. Don't go looking for additional information by hacking email accounts, asking other friends and digging the past. It's very dangerous and will engulf you in a negative environment.

Overcome sadness, let go

If you have failed your examinations, were not selected for a job, or a girl or boy rejected you for matrimony purpose or you have not got the grades you desired overcome the sadness. Let go what happened start anew. Don't be scared of failure, you are trying. Learn from your failure and make changes in yourself and your lifestyle. It's a journey, so go on.

Stop Judging

You have to stop judging yourself and others. You have to stop meaningless negative gossip and cut off negativity. All the judging can drain you of your mental capacity. Everyone is beautiful, lovable and valuable. Even you may be. Be kind to yourself and focus your energy where it gets you results. Empty judging never got anyone anything. It is easy to comment on someone else's hard work but very hard doing it. E.g. People always judge sports personalities ruthlessly. But do they care to see how many hours of hard work go to score better. Once we get sensitive to such things we will stop judging and start loving.

Stress and Positive Thinking

Stress in now so much a part of lives that is almost synonymous with work. Stress can be caused due to various reasons. Overworking at office, a tyrant for a boss, a troublesome wife, irritating relatives, naughty kids or parents who are always opposing whatever you do. Stress can affect you in different ways. But mostly it affects your minds, makes you tired, fatigued and you look like a droopy flower. Well with so much to do in life you cannot afford to go on like this. You can't escape the stress. But there is a way out to counter it. Positive thinking can be a great anti-dote to stress. A positive approach towards your stressors can help you deal with it very well.

Relax and take a break

Always take breaks when you feel overwhelmed with life. Just go on to a hill station with your spouse and enjoy nature and solitude. It will recharge you with a positive approach that is fresh and new. Sometimes just moving away from the hustle-bustle proves therapeutic. You deserve the break and do not feel guilty about it. You have worked hard and it's your turn to relax. Let your thoughts just be and feel the inner peace.

Calm down

In any panic situation it is very important to calm down. Calm down and think again. Think what you could do otherwise than simply panicking over what has happened. Choose to act differently. Choose to think positively. You will love the results because a calm positive mind is equipped to take better decisions. So when you are going through a lot of stress simply calm down instead of firing away at people or anyone on whom you could vent your frustration on. If it difficult to do so just moves away into a place where you are only with yourself and take deep breaths.

Nourish your body and soul

Your body is a temple. So it is important that you take care of it. Nourish it with healthy food and give it some nice pampering before you are off to bed. Your hard working body that even works while you sleep deserves a treat. Be good to your body. Don't look down in disguston your thighs or that paunch. Embrace yourself completely and promise yourself to make positive changes. Your body understands all your thoughts. So if you are giving out some nasty messages as you think about your body remembers it doesn't like it. Tell your body it's hardworking and beautiful.

Initiate prayer and engagein prayer, it nourishes your soul. Pray for peace, compassion, gratitude, for mercy and for all that your heart desires. Prayers make your soul happy and positive.

Meditate more

It is good to mediate at least for 20 minutes every day. It relaxes your mind, helps you see things that you do not see with open eyes. Meditation builds your inner world and connects you better with your outer world.

Breathe and be Positive

Check your breath. The way we breathe does not make full use of our lungs. You have to use you respiratory system to its best ability. Give it some exercises and take lungful of oxygen to refresh your mind and body. Breathe deep when you are getting negative thoughts. Breathe to calm down. Watch your breath. Breathing patterns like Pranayam help you cultivate positive thoughts in a calm mind and also beat stress.

Good Thoughts

From the time you were a small child you have been bombarded with thoughts. Thoughts that parents gave you, thoughts that teachers passed on to you and thoughts that friends shared with you all these thoughts influence your life and make you the very person you are today. Imagine of no one told you about ghosts would you ever even know of something called 'ghosts'.

It is only because your parents told you since you were a child that you must not lie you formed an idea that lying is a bad habit. Imagine why the world has so many terrorists. May be because someone is passing on a different thought to them on world peace they form a certain thinking leading to an action. Your thoughts have the power to come true. Therefore try having good thoughts and stop the negative ones at the right time. Here's how you can train yourself to think good thoughts.

Help yourself

You have to be kind to yourself. Forgive yourself and love yourself for what you are. You are just one of a kind, one in a million and have your own special space on this planet. So you have to start helping yourself to think good thoughts if you want this one in a million person to be a great one. Let go of negative thoughts. Give up thoughts of anger, hatred and discomfort.

Let them go and unburden you. Cup your hands and just think that all your negative thoughts and problems are in your hands. Now flung it out of the window or just throw them up in the air. Send them to the universe. It will be taken care of. Now you are ready to replace all those with new positive thoughts.

Heal up

Don't feel guilty about the past. It's gone. Forgive yourself and move on. Spend time with friends, do activities that give joy to you and work with commitment, without sabotaging your job and you will experience peace. Let you heal and recover. Start your day by remembering god and sorting out your thoughts. Accomplish the tasks you had decided on. Spend time with family, eat together and before you sleep, think what your heart desires. Meditate for some time and have a good night's sleep.

Good deeds

Work on how you could perform good deeds. Good deeds are not giving money in charity. Good deeds can be as simple as helping your mother or father. Help a neighbour who is sick. Help a child buy grocery. Help your friends when they need you. Talk to them often. Don't say you're too busy to call or message. Do a complete job at work and be honest to your employer. Good deeds begin with these small things that lead to the larger greater good. So decide which area you need to work on and start off, now!

You are the master of your mind

Always remember you are the master of your mind and don't let it work the other way. You can stop your thoughts, alter them, cultivate positive thoughts and even avoid thoughts that others give to you. Let yourself have a positive mind because every thought you think has the potential to come true. You have to decide what you want reality to be like. Therefore, it is said that you make your own destiny.

Positive Thinking Stories

Building your own positive thinking can seem difficult at first but as you start listening and reading to stories of people whose positive attitude have taken them far ahead in life you would like to believe in the power of positive thinking. Here are some true stories on the power of positive thinking that have results as good as magic.

Story of the boy who had no ears

Napoleon Hill in his book 'You can work your own Miracles' quotes this story about the power of positive thinking.

A man had a son who was born with no ears. He had no power to hear. But the father never believed this. He went on to treat his son the way a boy who could hear would be treated. He never made his son feel that he is deaf. He did not give up. He did not accept the fact that his son was deaf. He simply went on with his strong belief.

When the boy turned about ten years of age, he started hearing. When he was taken for a doctor's visit the doctor said that the nerves have somehow made their way to be in a form that supports hearing. The boy's body had responded to his thoughts. His father's thoughts influenced him to start listening. Nature showed its true beauty in allowing the boy to hear. This story captures the power of positive thinking that none other. Would you now like to believe everything you hear?

The girl who met the dolphins

There was a little girl who loved animals. Her dream was to see dolphins swimming. She loved the aquariums and zoos. Every day when she walked home she would see a big advertisement of the aquarium in Singapore. She would watch it and say one day I am coming to you. She had no means of travelling not even a passport. At once the situation turned such that she was forced to make a passport. She stood in long queues and hours of wait and lot of checks. Finally the passport arrived and she showed it at the required place of administration.

Within a few weeks she discovered she was chosen as an intern in an international company and had to be at her internship organisation away from her home. What country do you think she was being sent to? None other than Singapore! Her joy knew no bounds. She was at her destination even before she knew. She started work and at one of the weekends her colleague asked her if she wanted a little tour. She said a yes and off they went touring Singapore. They hit the aquarium and the girl was ecstatic. It was beautiful. Lots and lots of exotic sea life inside it, complete with a diver. But to her dismay no dolphins at all.

After that they went on to the zoo. Full with lovely animals the zoo was a treat. They next landed in an animal show and to the girl's delight there were dolphins that were performing in this show. They were swimming and bobbling and screeching. She loved the sight and thanked god. She had tears in her eyes. Her dream came true. Hold on, right after the dolphin show got over, in came some white Beluga whales. This was like a bonus for the girl. More than what she had asked for. The whales performed and even sang something that she only read in a book.

Be ready for success always be ready for success. Make plans and work on them. Think of all the possibilities. Dream big and have faith. Have good positive thoughts. You never know which might come true.

Self-Improvement

Well you might be finding ways and means to improve the way your life is now. That is because you may not be happy with your current life or maybe you desire to change some aspects of your life. For any change to occur in your life self-improvement is necessary. To begin self-improvement does not occur in a day. You have to commit yourself to the changes you want to make in life. The values you seek will have to be built as habits to see a real change. E.g. If you wish to lose weight but do not want to exercise daily it will not show any results. Similarly if you wish for high grades but do not prefer studying than the grades will only be what you do every day – a wish.

Value time

The most important thing for taking up any self-improvement exercise or value you must value time. It is the core of everything that you do. Never be late. Never procrastinate.

Avoid keeping people wait for even if it is your child or a close friend. Value time, be it yours or anyone else. You can do much better by just valuing time and reaping the benefits of doing things on time. Imagine how much time you would save if you prepared for work the night before. Keep the car keys in place. Make a mental note of breakfast.

Once you rise everything will be ready. You will glide smoothly in the morning and reach work on time. All of us know the results of procrastination so it is high time you did things you do not like before you do things that you enjoy.

Rise Early

Rising early, is the most important thing to make your life better. It is a war with your bed every morning. You may find it hard to leave the bed when you want to sleep five more minutes. Keep your alarm away from you so that you have to get up to stop it. Never go back to the bed once you have rise, simply head for your bath. Rising early gives you a chance to catch with the news. You can even exercise or have your breakfast in peace. Once you start doing it you will not find it difficult in some days. Rather you will like rising early for the energy it gives you and the great feeling that you are early riser compares to nothing else. Imagine how proud you feel when you rise early and finish important chores right in the morning.

Cut down on TV

Television is the biggest enemy in self-improvement. It is like a devil that keeps alluring and enticing you to watch it and leave everything else aside. Sometimes television turns out to be very expensive. They cost you your sleep, peace and your time. Keep a time slot for watching television. Avoid watching television late in the night. You will feel much better the next day at work.

Finish the work you have started

- Always finish what you have started. Leaving work before it is completed builds complacency and soon you will notice leads to procrastination. For all you know all your hard work, desires, wishes and expectation from your piece of creativity in gone in cold storage. So always finish what you start it will build your sense of accomplishment, boosting your confidence all the more to do things 100 percent.
- Stay away from negative people
- Choose to stay away from people who
- Make fun of you when you write your goals
- Laugh at you for rising early
- Shun you because you sleep early
- Forcing you to do things they do
- Don't think you are cool enough since you are irregular for late night parties
- Say you are mad when you share your dreams with them

Things around us influence us. Stay close to ice and it will make you feel cold. Stay close to fire and it will make you feel hot. Similarly people around you can influence you by sharing their thoughts with you. So always stay with people who are positive, who respect you for what you are and encourage you in your endeavours. Sometimes a few words of encouragement and a belief from someone else in us may lead us to do great things. Having the right people around us is very important for self-improvement.

Optimism

The example of looking at a glass half filled with water as a glass half filled or a glass half empty is a popular way of showing what optimism is. Optimism is looking at situations and problems with a positive attitude. The glass is half filled is a view of someone who is positive and is seeing the presence of water that can quench thirst. Whereas a negative person will only see the glass half empty and be dissatisfied. Having an optimistic vision always gives you the strength to look beyond and find ways to rise above the problem. Here are some techniques that can help you be optimistic.

No Cribbing

Stop cribbing. My hair is too short, I am overweight, Why I don't have enough money, I can't eat so much, why am I not as beautiful as Aishwarya Rai, why am I so poor and why do I have to suffer with the drudgery of everyday life? These are questions and statements we make very often. So very often we are only cribbing. But do we really see what we have got in us that make us unique. A healthy normal body is a blessing that many would trade off with all their money, a heart that is beating when someone is fighting for life. You have a comfortable bed to sleep in while many sleep on the roads. Feeling ant different? We are blessed with so many blessing that many others would have been robbed off. So being grateful and accepting yourself is the beginning of optimism.

No comparing

Comparing yourself with others can make you feel worse. Never compare because you might not know what troubles the person must be going though and would love to be in your place. Avoid being in awe of materialistic possessions of others. If someone else has a diamond ring it's not necessary that you too need to have one. Never compete. It only makes you bitter. Compete with yourself. Try moving to the next level while you are racing against yourself and no one else.

Look at the brighter side

Always look at the brighter side. Thank god that you were not born in a land that is facing wars. Be grateful that you have eyes to see the beautiful world and hands that help you every day. You have almost everything that you need. You have clothes, a house, the best of foods, entertainment and everything else to keep you happy. If you get low scores on your math test, think that at least you have marks on your paper. You do not have a zero. That means you have some amount of knowledge and with practice you can get better.

Look for Solutions

Always look for solutions than drowsing yourself in sadness and depression. The little story of the thirsty crow you learnt in school serves as a good example. The crow was thirsty and was looking for water. He found water but could not drink it. Instead of getting angry, stamping his feet and saying rude words he thought of a solution and soon had a nice drink of water and flew away. The crow was smart to look at the brighter side and create an innovative solution with his positive thoughts that saved his life. The crow was optimistic and his optimism got him a bright solution.

A Scientific Perspective on Personality Development

Some people whom I talk about my work with MET / EFT press the knock therapy like the stamp "esoteric nonsense" on. The obvious and far-reaching effects that can be achieved here regularly seem to be-without damn scientific evidence or explanations "is not possible". So I thought to myself, I explain the impact model of MET /EFT and the field of personality development times from a scientific standpoint.

Perception of reality

According to Dr. Dispenza we create ourselves our reality at the moment, as we reflect on the world. We're changing our brains about how we think. Here are the basis past experiences and associated feelings.

Our memory is a process of maintaining new synaptic connections.

Personal Change

We interpret reality constantly, based on our educated patterns and perceptions. Here, no one likes to change, unless he recognizes it as a necessity. The thoughts and feelings that determine a person's life solidify itself as a biofeedback loop and are therefore only to break through pain, because a change causes painful body chemical reactions.

To change this setting to life step by step, it is necessary initially to the recurring thoughts to be aware of the particular harmful. This attention is an art that must be practiced constantly. You'll find the surprise that most inner attitudes do not have to be true. The unconscious thoughts were all made by automated and were thought were finally aware of, such as the activity of driving a car. These unconscious thought patterns have as much an effect on our daily lives as conscious thoughts. "The more we are exposed to the same stimuli, the stronger we anchor ourselves to the corresponding objects in the external world."

The brain and the body also strive not to waste energy and do not require the use of familiar ideas. That is why it is so easy to be like we are.

We live on the principle so forth as we taught and how we were brought up and it is when we become aware and actively change something, we can live as we want.

Influence of our family heritage

Synaptic patterns are passed on, are limited not only on factors such as height and hair color, but also on attitudes and inclinations. In short, we take the emotions and tensions of our ancestors in us. That is, if we are genetically e.g. by our parents in childhood on negative emotions like a victim attitude or <a a

weak self-confidence could also be programmed, trends were to eventually even our unconscious thought patterns and We have our whole lives with a victim attitude and low self-esteem, if we do not change. We act according to our pre-programming.

Statements are fully in line with my experience. There are certain patterns moves through generations of families, such as Childlessness or certain medical conditions. Just as the behavior of many people is very similar to those of their parents, even though many hear the hate

It is your choice to feel joy or anger.

Feelings

Emotions are chemical memories. Your current situation your rating because of feelings from the past.

So you want to change your experiences, you can change your feelings. And then also your life will change for the better. Luckily, there are MET & EFT techniques nowadays to just alone to achieve sustainable and anytime. Fantastic, is not it?

Stress

Living under stress is to live in survival mode – is virtually one and the same. We are constantly under stress or are in front of the hat, we always solve the alarm functions of our body. He then has no time to regenerate.

The consequences of long-term stress are numerous and catastrophic, here is a list: Chronic fatigue, depression, lack of motivation, sleep disorder, disease susceptibility, a weak sex drive, concentration problems, great need for routine, irritability, heart problems, digestive problems, muscle pain, back pain, obesity, high cholesterol and blood sugar problems.

Only if we are able to turn off the thoughts that trigger the stress response, the body in the ideal case has the power to begin the healing.

Quite a lot of physical symptoms for the causes of stress no wonder our society is getting sicker.

Condemnation

The chemical stability of each emotional state, and we employ every day for years, has an effect on us. Now we can better understand what is meant by the old saying,

Every judgment falls on us back

Exactly what we condemn in others, we also condemn us forever.

Emotional addiction

Are we looking at here no external stress triggers are available, we organize us. And we do not find, and then we will produce it, physically or mentally. The voice within us is the voice of our body that wants his old order and gets rid of his discomfort. In an ideal world, we would realize our addiction to certain emotions before they cause us real damage. However, it is more common that most people only come through physical stress symptoms figure out what's going on with their emotions.

I think everyone knows people who always complain about just what and suspend. These people are simply addicted to the emotion anger. Unfortunately, this species is very strong in Germany. Conversely, the statement that we are also addicted to pleasure, and this is then always re-established all a matter of habit or emotion to which we are now addicted.

The role of the body

If we cannot think beyond our own emotional state beyond, we live the way our environment dictates our body. We can only do through our Spirit and change a decision to conscious life

A person responds to a relatively benign disorder with a violent emotional outburst, then these are the loud and persistent signals that sends his body. Chemicals move through his body and the brain, the autonomic nervous system has seized the control itself to meet the needs of the body are. In that case, the body, and not the mind, control of our behavior. The less aware of our thoughts are, the stronger the body dominated our behavior. This process is to break only through conscious awareness.

The good news is: We can train to be healthy and happy instead of sick and depressed.

This reminds me very much of the feeling of a Wutausbruchs. That's exactly the point, we are so overwhelmed by our emotions, which at the moment we cannot act otherwise. But luckily there is a possibility with MET /EFT after the überbrodelnde emotion or better already weakening.

Get Support

Without expert assistance projects go as removing or other personal projects often wrong. The experienced eyes and ears of an expert or outsiders can see detail when something went wrong and how to correct it.

Which I can only agree with I even have built up a network of mentors, trainers, coaches and peers for my personal development.

The brain does not care

The brain does not distinguish between positive or negative feelings. One does not make him more trouble than the other. So we have to use the brain to realize a new way, rather than us to be tempted by the entertainment media or our environment to thinking in predictable ways.

Conclusion

The author, Dr. Joe Dispenza has presented several scientific sources of the latest findings in brain research fully and articulately in his book. The knowledge I have in my article covers only a small part of it by reference, are surprisingly congruent with the alternativen/spirituellen- whatever-of explanation of personality development. In addition to the attempts to explain how we will measure our reality in the book also explains more practically how each of us can influence themselves.

Personality development has always been a key concern for researches in personality. But whether personality traits change in a meaningful way during adulthood and when those changes takes place has always been a conflicting issue.

Available research related to development in personality is based on three pioneer views proposed by luminaries of psychology. In his view, Freud stated that personality was fixed by the age of 5, at the completion of the oedipal stage of development. Later, Erikson proposed that development continued throughout the life span, with the most disorderly time occurring in adolescence. James, proposed that personality develops to a point, and then is effectively 'set like plaster' by the age of 30. Each of these theories has been influential in forming the ways that scientist and policy-makers view personality development, psychopathology, public policy and more fundamentally our worldviews with respect to human nature. Of early theorists, the perspectives of Freud, Erikson and James represent the full spectrum on personality change, from very little to potentially very much.

10

Mind/Body Connection: How Your Emotions Affect Your Health

What is good emotional health?

People who have good emotional health are aware of their thoughts, feelings and behaviors. They have learned healthy ways to cope with the stress and problems that are a normal part of life. They feel good about themselves and have healthy relationships.

However, many things that happen in your life can disrupt your emotional health and lead to strong feelings of sadness, stress or anxiety. These things include:

- Being laid off from your job
- Having a child leave or return home
- Dealing with the death of a loved one
- Getting divorced or married
- Suffering an illness or an injury
- Getting a job promotion
- Experiencing money problems
- Moving to a new home
- Having a baby

"Good" changes can be just as stressful as "bad" changes.

How can my emotions affect my health?

Your body responds to the way you think, feel and act. This is often called the "mind/body connection." When you are stressed, anxious or upset, your body tries to tell you that something isn't right. High blood pressure or a stomach ulcer might develop after a particularly stressful event, such as the death of a loved one. The following can be physical signs that your emotional health is out of balance:

- Back pain
- Change in appetite
- Chest pain
- Constipation or diarrhea
- Dry mouth
- Extreme tiredness
- General aches and pains
- Headaches
- High blood pressure
- Insomnia
- Lightheadedness
- Palpitations
- Sexual problems
- Shortness of breath
- Stiff neck
- Sweating
- Upset stomach
- Weight gain or loss

Poor emotional health can weaken your body's immune system, making you more likely to get colds and other infections during emotionally difficult times. Also, when you are feeling stressed, anxious or upset, you may not take care of your health as well as you should. You may not feel like exercising, eating nutritious foods or taking medicine that your doctor prescribes.

Abuse of alcohol, tobacco or other drugs may also be a sign of poor emotional health.

Why does my doctor need to know about my emotions?

You may not be used to talking to your doctor about your feelings or problems in your personal life. But remember, he or she can't always tell that you're feeling stressed, anxious or upset just by looking at you. It's important to be honest with your doctor if you are having these feelings.

First, he or she will need to make sure that other health problems aren't causing your physical symptoms. If your symptoms aren't caused by other health problems, you and your doctor can address the emotional causes of your symptoms. Your doctor may suggest ways to treat your physical symptoms while you work together to improve your emotional health.

If your negative feelings don't go away and are so strong that they keep you from enjoying life, it's especially important for you to talk to your doctor. You may have what doctors call "major depression." Depression is a medical illness that can be treated with individualized counseling, medicine or with both.

How can I improve my emotional health?

First, try to recognize your emotions and understand why you are having them. Sorting out the causes of sadness, stress and anxiety in your life can help you manage your emotional health. The following are some other helpful tips.

Express your feelings in appropriate ways. If feelings of stress, sadness or anxiety are causing physical problems, keeping these feelings inside can make you feel worse. It's OK to let your loved ones know when something is bothering you. However, keep in mind that your family and friends may not be able to help you deal with your feelings appropriately. At these times, ask someone outside the situation--such as your family doctor, a counselor or a religious advisor--for advice and support to help you improve your emotional health.

Live a balanced life. Try not to obsess about the problems at work, school or home that lead to negative feelings. This doesn't mean you have to pretend to be happy when you feel stressed, anxious or upset. It's important to deal with these negative feelings, but try to focus on the positive things in your life too. You may want to use a journal to keep track of things that make you feel happy or peaceful. Some research has shown that having a positive outlook can improve your quality of life and give your health a boost. You may also need to find ways to let go of some things in your life that make you feel stressed and overwhelmed. Make time for things you enjoy.

Develop resilience. People with resilience are able to cope with stress in a healthy way. Resilience can be learned and strengthened with different strategies. These include having social support, keeping a positive view of you, accepting change and keeping things in perspective.

Calm your mind and body. Relaxation methods, such as meditation, are useful ways to bring your emotions into balance. Meditation is a form of guided thought. It can take many forms. You may do it by exercising, stretching or breathing deeply. Ask your family doctor for advice about relaxation methods.

Take care of yourself. To have good emotional health, it's important to take care of your body by having a regular routine for eating healthy meals, getting enough sleep and exercising to relieve pent-up tension. Avoid overeating and don't abuse drugs or alcohol. Using drugs or alcohol just causes other problems, such as family and health problems.

The Psychology behind Overweight

Health and weight is intimately connected with your

psychological, emotional, and spiritual state. Overweight can be a symptom of many different conditions. You might have used excess weight as protection, like insulation. Maybe you have used weight as a way of not allowing yourself to be as beautiful and attractive as you are because you do not desire the attention you might get from people. Or you might have used it as a way of abusing yourself, not believing that you deserve love, attention, affection and care. Remember, these beliefs are just that - beliefs. You have been conditioned to belief this, it does not by any means indicate that this is in fact the truth! Keep this in mind, at all times. If you are ready for it, ask yourself why you are overweight and in less than optimal health? Set aside some time to meditate upon this.

With every pound that comes off, there will be a psychological effort and effect as well. Every pound comes with a psychological condition, hang up, or belief. If you are not ready to let go of that particular condition, the weight will not come off. It will come off easily, but will pile on as soon as it is off. If you are not ready to let go off the psychological condition, your weight will not change long term. You might have discovered during your meditations how there might be some resistance within you to let go of some of your poor health habits, excess weight, and old beliefs. That is why it is so important that you give yourself enough time and patience. It is a process and it will take the time it takes. If you are not ready to tackle the underlining reason for your weight and health condition, your condition will not change. This does not have to be a conscious process, just allow for issues to come up as they do.

There is no reason to force you. In fact, you are probably better of releasing some of the pressure you are already putting on yourself. Along with your improved state of wellbeing, these issues will surface whenever it is the right time. It is an unconscious process and demands no conscious effort, so be compassionate with yourself.

Keep in mind that you do the best you can with the tools that you have at present. Increase your tools and you might be able to handle the situation differently.

11

PERSONALITY AND MASKS

Personality Change due to Environment

Screams, Widening eyes Stares, Locked doors. Then suddenly silence. Dreary silence that causes a bit of panic in you, Your eyes dart left and right, looking at the people that were at one point screaming, but only stare in silence now. Some are looking at other people. Some are looking at you. Their stares are deafening; you only want to leave this place. You're paralyzed with fear. Then you leave this locked place, solitude of what looks like insanity. Your disposition changes, your fear are relieved, and suddenly you're into another section of the building. You have just left the Sheltered Freedom unit, a place for aging people with dementia, at Bessie Burton Sullivan Skilled Nursing Residence. It's unlike most nursing home environments; the doors are locked, people are wandering around talking to no one, and the residents at times burst into agitated screams. Yet one floor higher, you have what looks to be your normal nursing home. People in wheel chairs some hard of hearing, others talking or watching television with their roommate. Each person, with a different personality, may be that way because of the surrounding environment they live in. Would one burst into agitated screams if placed into a room of seemly non-dementia residents? But the grander question is, is the personality of nurses and visitors greatly affected by the people and the environment they spend time around? Or is it simply a mask, used only in the situation of the nursing home? The environment causes one to use a mask that would otherwise go unused in outside social situations; it is not a change in personality, but a change in disposition that lasts not only for the time being spent in the given environment, but also makes itself available for use in other environments that may be like that of the original environment.

To say that our personality, our "self", changes depending on our environment could in fact be true. Gergen points out that "the centered self begins to collapse under the demands of multiple audiences". With the advent of just another audience (that being the nursing home residents and staff), it could be said that it is contributing to the loss of self, and the creation of a decentered persona. But by making that statement, I am implying that any new group or person that is encountered would be leading to the destruction of the self. That is quite a statement to make, without further exploring what is meant by "self." Gergen points out that "modernist views of the self dominate the profession of psychology". The argument goes that we are nothing more than input/output machines. Of course, you have the Galen Straw son definition: "specifically, a mental presence; a mental someone; a single mental thing that is a conscious subject of experience, that has a certain character or personality, and that is in some sense distinct from all its particular experiences, thoughts, and so on, and indeed from all other things" . Either of these definitions of the self still led me to believe that the self is being deconstructed by the environment it is in. Masks are being created or acquired, and the centered self that one might have once thought existed, may cease to exist, or become buried among the many masks and images that are now projected to the audiences encountered. But even with a modernists view and a philosophical materialists' view of the self, it does not and cannot give us a clear answer as to whether the self is being deconstructed or not. It only leads us to two contrasting views of what self may in fact be at a given time in a given person.

Since we cannot make adequate judgment on a given persons' self, or personality at this time, we have to use the next best thing. Observation of not only the residents at Bessie Burton, but also staff and volunteers interaction with them, as well as a self analysis, should give a clearer picture that the environment does in fact effect a person's personality.

Observations of staff interaction with residents provide a good example. On the first day of working with my new "audience" I noticed that many nurses did in fact talk with residents. Some were rather short conversations about such little things as the weather outside; others were about specific things occurring in the lives of residents. One such circumstance comes to mind during the reading of the morning news. While talking about a given event in the news that morning, a tangent came in which the conversation had turned to the ill health of one of the residents roommates. While the difference in personality of the staff member was very apparent in the sound of her voice, as well as mannerisms, the other residents in the room had little expression change. They sat and said nothing, while the conversation between the staff member and the resident continued. While the resident, who I had been talking to no more than five minutes prior, had seemed to be in good spirits, now seemed a totally different person.

Analysis of the above story has some interesting points to be made about the observations. The staff member, who I had been in contact with for more than 2 weeks, had done a complete turnaround in personality when with residents, and was completely different when not around them. Where as she were talking to children when talking with older residents, the moment she wasn't around them, a much stronger person appeared. I attribute this to a mask, and not a personality change; the change was only a change in disposition for the moment and place, not necessarily for other environments. "Each movement of the body, seemingly private and spontaneous, is orchestrated for social effect," Gergen claimed. The resident on the other hand, seemed to be reflecting inward, and not showing a mask. Although the change was very adverse from his original character not 5 minutes prior, the situation was different for him then the staff member. Whereas he was directly affected by his roommate's ill health, the staff member was not. The direct affect of the environment I think has a major impact on personality; whereas Gergen talks of multiple audiences, which could affect the staff member, whose life there at the nursing residence is work, and has a life outside of other people they may know. While the resident, who lives at this place, doesn't necessarily have the same scope of audiences in which to use multiple masks, and therefore would have a more centered self. This centered self, which has probably changed over the years to accommodate new surroundings, and to accommodate new people, may not be developing the masks that younger generations now use more often than not. Gergen states that "each of the selves that we acquire from others can contribute to inner dialogues, private discussions with ourselves..." This could be the case for the resident; he may not have been using a mask, but instead reflecting inward on one of the many personas, selves that had been taken over the years.

Turning the analysis from outside sources such as residents and staff to a Freud-like look into one's thoughts is at once frightening, yet intriguing. It would be easy for me to say, no, that my personality is not changing, and that it is still the same as it was when I begin to work there. But that would not be looking into what I have seen and the changes that have occurred. Starting out there, I had little to no knowledge of the inner workings of a nursing residence. I had no prior experience in working with patients who experienced dementia. Most people would find this as a hindrance, but in fact, it was actually useful in giving a clear view of any masks that would occur that had not been used previously. Immediately upon beginning work there, talking with people had changed. My rate of speech had slowed, my diction had become better, and my attitude had changed from that of working with adults to working with almost children. At first, I caught myself doing this; I had figured I was being stereotypical, but was in fact only mimicking what I had heard and seen others doing there. Gergen points out that "all selves lie latent, and under the right conditions may spring to life" (Gergen, 1991, p. 71). It could be that I had this was not being stereotypical, but in fact drawing out a mask that I had not used, but had absorbed at an earlier time, that had suddenly "sprung to life" by the environment. Upon leaving the floor, all changes that had occurred, were now non-existent. I see this as use of a mask in a changing environment. I soon observed other volunteers doing the same thing; while they acted on way with residents, and spoke in a certain matter, they changed upon leaving the floor, and even more by leaving the residence. I do not attribute the change in personality to be necessarily a change in the core personality of a person, but yet only the creation of a mask, that they may or may not know they are using. While looking at myself, I know that I am using the particular mask when working at the nursing residence. But I also understand that I use a number of different masks a day, "in meeting the demands of everyday life" and to simply live.

We do not change personality, but our disposition changes along with what the environment elicits, therefore causing a mask to be brought to life, or drawn out from a multitude of selves, which is available for that given time and situation. It could be argued that the environment that we are in is not affecting our personality, but that we are effecting the environment, and therefore eliciting a response from that environment, therefore causing us to create or draw out a new mask, or new idea which is worked into our personality. The idea that one cannot simply be manipulated by the environment around them, but instead will change something in that very same environment, that may then affect them, is not a new idea.

Gergen points out in The Saturated Self of reporters putting their point of view on what we take as factual news. But even then, can we still say that environment is causing an effect in our personality? Was it not Hume who said "these question when he was talking about self Hume, could the same be said about environment affecting the self. It is helpful to keep in mind that Hume also said that "but there is no impression constant and invariable" which is what we're actually looking at here. Personality based on environment is a constantly changing phenomena; Gergen points to it as the fall of personality, with people being only identified by the masks that they substantially create. People will continue to create new masks for different environments, audiences, and people, no matter what the circumstance may in fact be. Even in this paper it is present; in my bid to write an objective paper on the environments affect on personality, not saying whether it is good or bad to have these masks, I have placed my own spin on the facts. I have leaned towards saying that the decentered persona is the better of the two; allowing much more communication between people to occur. The centered self, although may be present, is being left in the past, being converted to many interchangeable masks, to better adapt to the ever changing conditions around us. The environment causes us to use masks; and depending on the environment, also dictates how long that mask is used, and when it should be used again.

Impact of Socio-cultural Factors on the Personality Development

Children born and live not only in a society but also in a specific part of it, and are therefore, influenced by particular subcultures of class, race, religion, and region, as well as by specific groups such as family and friends. During their lifetime, they continually encounter new or changing conditions, both personal and social, and must learn to adjust to them. The most important socialization, however, occurs during infancy and childhood, when the foundations of later personality are laid. Whether a child becomes outgoing or shy, intellectually advanced or average, or energetic or subdued depends on many unique influences effects are difficult to predict at the child's birth. A variety of factors influence child development. Heredity guides every aspect of physical, cognitive, social, emotional, and personality development. Family members,' peer groups, the school environment, and the community influence how children think, socialize, and become self-aware. In this research a cross sectional study was conducted in Faisalabad city The purpose was to study the impact of socio-cultural factors on the personality development of adolescents.

The study was confined to children of age 11-16 years. Mothers were interviewed through a well-structured questionnaire consisting of open ended and close-ended questions. The sample size was 600 respondents. Faisalabad city was divided into three categories on the basis of upper, middle, and lower income groups by the Federal Bureau of Statistics and according to this criterion two localities from each category was selected randomly. Moreover focus group discussion from 36 persons representing different areas and income groups was also made to make their views part of this study.

This study entails with the major socio-cultural factors that affect the children personality. After a comprehensive research the factors which are identified as important determinants of child personality are parent's economic conditions, their education level, family size, family structure, children's socialization level, peer group, school environment, parent's involvement level in children's day to day activities, extracurricular activities and government policies. On the basis of results and findings of the study, different recommendations are proposed; focusing on parents' reasonability towards child education, governments role in providing equal quality education to all classes of children compared emphasizing on improving parent, teacher and children relationships. This study also highlighted and suggested areas of further research. Subject having potential to be studied individually. Moreover, the role of media and other new emerging sources of mass communication should also be investigated. More studies on children's psychology gives understanding to better personality development of children. Nurturing children in the best way is the key to attain broader objective of building an educated, well behaved society and nation and this objective can be achieved through conducting research on children psychology.

Personality and sociocultural development during early childhood

This chapter looks at the major perspectives that form the basis for personality and sociocultural development during early childhood. Topics covered include coping patterns, aggression, prosocial behavior, the effects of peer interactions, and continuing development of self. Children learn to manage a wide range of feelings and emotions. The important to emotional development is the child's ability to cope with fear and anxiety. Fear is a response to a specific situation and anxiety is a generalized emotional state. A child may experience regular and continuous feelings of unease, often without knowing why. Children can be help cope with fear and anxiety by parents reducing unnecessary stress, being a role model, seeking professional help.

Children can also cope with fear and anxiety using defense mechanism such as identification, projection, denial, reaction formation, displacement, regression, rationalization, repression, and withdrawal. Children are expected to inhibit the display of some emotions such as anger, distress, affection, joy, sensuality and sexual curiosity. Children experience developmental conflicts their needs to depend on their parents and their desire of independence dealing with compliance, mastery and competence which Erikson identified as autonomy verse shame and initiative verse guilt.

According to Erikson, children either become more independent and autonomous if their parents encourage exploration and freedom or they experience shame and self-doubt if they are restricted and overprotected. In addition, children view of them undergoes major change as they face conflicts between the desire to act independently of their parents and the guilt that comes from the unintended consequences of their actions. Parents who react positively can help their children avoid experiencing guilt. As children develop, their play becomes more social and engage in social pretend play involving the use of imagination, sharing of fantasies, and the inclusion of agreed upon rules. This help children deal with fears, provide companionship during periods of loneliness, and provide reassurance. Research indicates that 65% of young children have imaginary companions. Imaginary companions help children social skills and practice conversations. Children who are adept at imagination may be better at mastering symbolic representation in the real world. Children who are rejected by their peers in early childhood are likely to be rejected in middle childhood as well. They are also more likely to have adjusting problems in adolescence and adulthood. Rejected children may be aggressive or withdrawn and may be out of sync with their peers' activities and social interaction.

Children learn to incorporate the values and morals of their society into their understanding of themselves through internalization. Children develop a self-concept, their identity, or their set of beliefs. It is seen Young children are tend to describe themselves in terms of their physical characteristics, possessions, or activities. The tendency to describe themselves in terms of social connections increases. Children tend to imitate their parents. Children self esteem enhance by parent praising their children, encouraging and giving them responsibilities, allowing them to explore their potential freely, and showing them unconditional love. The sense of being a male or female is well established by the time children reach the preschool years. Children learn gender-related behavior and expectations from their observation of others' behavior as well as from books, media, and TV. Parents play an especially important role in the development of young children, particularly with respect to how parents exert control and express warmth. Authoritarian parents tend to produce children who are withdrawn, fearful, dependent, moody, unassertive and irritable. Permissive parents tend to produce children who are rebellious, aggressive, self-indulgent, socially inept, creative and outgoing.

Authoritative parents tend to produce children who are self-reliant, self-controlled, and socially competent with high self-esteem and do better in school. Indifferent parents tend to produce children who are free to give in to the most destructive impulses. How parents manage discipline is an important aspect of the effect that parents have on their children's development. The aim of discipline is not only to control children behavior but also to help them develop emotional self control.

Middle childhood is the span of years from age 6 to 12. At age 9, growth spurt for girls and 11 year olds for boys. Growth is influenced by activity level, exercise, nutrition, gender, and genetic factors. Gross motor skills such as running, jumping, and hopping and fine motor skills continue to develop and improve. Children begin to develop interest in sports. During middle childhood, children in developed world receive good nutrition so most height and weight differences among children are due to genetically determined factors. Children in developing world grow smaller than their counter parts in affluent advance world.

Obesity is defined as body weight that is more than 20% above the average for a person of a given height and weight. In United States, about 17% of children are obese. Most of children who are obese continue to be seriously overweight as adults. Obesity leads to high blood pressure, diabetes, and other medical problems. The cause of obesity can be a genetic factor, environmental factors, television viewing, lack of exercise and parental encouragement. The leading cause of death in middle childhood is accidents and associated injuries.

Psychological disorders and mental illness can begin in middle childhood, raising concerns about accurate diagnosis and treatment such as autism and attention deficit and hyperactivity disorder. Piaget referred to middle childhood as a period of concrete operations and encourages the use of concrete objects for teaching such as blocks, rods and seeds. Piaget stress that teaching should be through showing rather than telling because children learn by doing and they are active learners who construct their own theories about how the world operates. During middle childhood, short-term memory capacity improves significantly and understandings about the processes that underlie memory emerge and improve during middle childhood. Children's memory strategies and techniques enhance with age and develop the process of monitoring their own thinking. When children attend school, school teach facts or concepts, give directions for a particular lesson, state general rules of behavior, correct, discipline and praise children and introduce children in other miscellaneous activities. Children learn more in classes in which time on task is maximized, in which the teacher spends at least half the time on actual teaching and less on such concerns as maintaining order. The main emphasis on school are teaching learning and thinking skills, tailoring instruction to the child's individual learning style and developmental level, and fostering independent, self-regulated, self paced learning, learning in small groups and cooperative rather than competitive learning.

School success is influenced by many factors including achievement motivation which is an acquired culturally based drive, gender, and parents of successful children who have realistic beliefs about their children, have high expectations, are authoritative parents and talk to, listen to, and read to their children. Developmental and intellectual disabilities such as mental Retardation, depression, attention deficit disorder, and learning disabilities children and other special needs children all have afforded educational opportunities in least restrictive environment.

During middle childhood, according to Erikson, the central task focuses on industry versus inferiority. Children at this stage are characterized by a focus on efforts to attain competence in meeting the challenges related to parents, peers, school and other complexities of the modern world. Children self concept and self esteem continue to develop. The development of self esteem is a reciprocal process. Parents can positively influence their children's self esteem by offering realistic praise and by encouraging them toward activities in which they can be successful. Children use social comparison to themselves to abilities, expertise, and opinions of others. When objective measures are absent children rely on social reality such as how others act, think, feel, and view the world. In middle childhood, most friends are of the same gender; and friendships during middle childhood serve many functions. Peer relationships provide emotional support and help kids to handle stress, teach children how to manage and control their emotions, teach about communication with others, foster intellectual growth and allow children to practice relationship skills.

According to Selman, friendships develop through four stages: as playmates, then awareness of another's feelings emerges, then trust develops and finally children can look at relationship from another's perspective. Prejudice is a negative attitude formed without adequate reason which is directed at a defined group of people. As children grow older, they become capable of thinking with greater complexity and prejudice can be reduce by enhancing through cooperative activities that are important to children and promoting equality and disconfirm negative stereotypes. Popular children are helpful and cooperative, have a good sense of humor and emotional understanding, ask for help when necessary, not overly reliant on others, adaptive to social situations, and social problem solving skill competence. Unpopular children lack social competence, are immature, are overly aggressive and overbearing, withdrawn or shy, and are unattractive, handicapped, obese, or slow academically. Several programs teach children set of social skills that underlie general social competence. Although peers become very important in to children in middle childhood, the family continues to be children most important socializing force. Effective parenting in middle childhood can involve in increasing children's social competence through encouraging social interaction, teaching listening skills to children, making children aware that people display emotions and moods nonverbally, teaching conversational skills, including the importance of asking questions and self-disclosure and not asking children to choose teams or groups publicly.

In addition to other changes, children experience in early relationships between siblings can shape how children relate to others and choices made in later life. Also, in most cases, children fare quite well when parents are loving, are sensitive to their children's needs, provide appropriate substitute care, and are good adjustment of their children. When parent divorce, children are most likely to exhibit behavioral difficulties, anxiety, depression, and low self esteem and they often have more problems with school. School-age children tend to blame themselves for the breakup. Twice as many children of divorced parents require psychological counseling as do children from intact families. For some children, living in a home with unhappy marriage and which is high in conflict has stronger negative consequences than divorce. Blended families include remarried couple that has at least one stepchild living with them. Living in blended family involves role ambiguity, in which roles and expectations are unclear.

Adolescence is a remarkable time of growth and development; in just a few years, children transition dramatically towards adulthood across multiple domains. Adolescence is physically the healthiest period of the lifespan. There is a dramatic body parts grow at different rates due to higher levels of testosterone boys experience greater increases in muscle growth; girls experience an increase in body fat. Body shape differentiates as boys develop wider shoulders and girls develop wider hips. These biological, as well as cultural, factors can cause depression and anxiety in females at this age; an emphasis on exercise may help to keep females active and combat negative self-images. Primary sex characteristics include sex organs in males and females grow significantly to allow room for sperm and egg production. Secondary Sex Characteristic for both male and female includes growth of pubic hair, underarm hair, facial hair, and arm and leg hair. Skin becomes rougher and oilier, bones become harder, the voice becomes lower, and the chin, nose, and ears become more pronounced.

Hormones have a powerful effect on the brain, influencing its development. However, the emotionally often seen in teenagers results not only because a hormone action but also because of complex sociocultural and environmental factors. Adolescents have the ability to begin moving from childhood toward adulthood due to their cognitive development. This is the ability of the brain to begin processing more abstract thoughts. Some of these thoughts, indeed many of these thoughts, are focused on themselves. By being able to think abstractly, which is a new developmental ability? Now, as adolescents, the journey toward self-reflection and self-identity, may begin. By asking clear self-identity questions, they may find answers that will be enlightening, even insightful and complex. They will strive to learn to make good choices and decisions toward their future as a responsible citizen. This process is often difficult for adolescents. They may change periodically in terms of their self-concept. According to Piaget, adolescent gain the ability to think about intangible objects and methods and have the ability to see multiple aspects of one idea. As adolescents enhance their understanding of themselves, they actually become more aware of their own emotions and feelings and how these feelings affect their daily lives. By gaining some emotional understanding of themselves, they are able to change their self-identity. This is how they perceive their characteristics and abilities fit with the opportunities that are available to them. These changes are now known to continue in our American society well into emerging adulthood. But many of the identity issues that begin during adolescence determine the paths an adolescent may take including future college, vocational or career choices, as well as other aspects of their lives.

Adolescent constantly views themselves as the center of attention and certainty of an individual's distinct personal experience and fate. It is here clear that adolescence also belief that unfortunate occurrences only happen to other people which encourages risky behavior. Adolescence today continue to be highly sexually active and about 20% of sexually active teenagers have sexually transmitted disease. Teenage mothers and fathers are associated with difficult economic circumstances and personal challenges. Marriage under such circumstances generally does not produce positive outcomes in part because early marriage often leads to dropping out of school.

During adolescence, young people reach physical maturity, develop a more sophisticated understanding of roles and relationships, and acquire and refine skills needed for successfully performing adult work and family roles. The developmental tasks of this period--coping with physical changes and emerging sexuality, developing interpersonal skills for opposite-sex relationships, acquiring education and training for adult work roles, becoming emotionally and behaviorally autonomous, resolving identity issues, and acquiring a set of values are all tied to successful functioning in adulthood in one way or another. The movement toward adulthood colors our expectations of adolescents, and hence our treatment of them. One expect adolescents to move away from the adult-directed activities of childhood toward the autonomy, responsibility, self-direction and independence from their parents and forming an identity.

Consistent with these expectations, adolescents are granted increased freedom of choice to varying degrees; adolescents select their academic courses, choose their friends and activities, and make plans concerning post high school education, employment, and family life. Many of these decisions have important implications for young people's subsequent life course.

We may say that educational decisions, such as whether to attend college or not, affect future career opportunities and vocational development. Similarly, becoming an adolescent parent often limits educational attainment and employment opportunities. Erikson viewed the critical developmental task of adolescence as identity verses identity confusion which requires the teen to sort through various choices in order to answer to questions "who am I?" Adolescents who go out with friends rather than study for an important test, who engage in unprotected intercourse or experiment with a new drug, or who ride home with an intoxicated driver may unknowingly affect the direction of their future lives. Moreover, short-term choices may evolve into regular patterns of behavior or lifestyles, which, in turn, influence future development. Thus, the choices that adolescents make and the developmental course they define can profoundly shape their later lives. Therefore, the adolescent's movement toward autonomy entails both growth and risk. On the one hand, adolescents need to experience greater freedom of choice so they can begin to exercise self-direction.

Successful parents must provide support to teenage children. Maintaining communication helps reduce serious conflict.

Parental monitoring is based on open communication and adolescent willingness to disclose the details of the adolescent's life. The most importance of peers increases enormously during adolescence. Through social comparisons, teens compare themselves to their peers as a means of defining themselves. Early on, dating serves to give young adolescents experience without deep emotional involvement. Later, adolescents who date may develop emotional closeness and serious romantic relationship. Adolescence and emerging adulthood is often characterized by risk taking behavior. Because the brain region related to judgment and emotional control are still developing, adolescent may take risks without fully appreciating the consequences. Suicide is the third leading cause of death during adolescence, and the rate of suicide in this age group is rising. To support positive adolescent development, we should support and strengthen families, provide then with activities in which they can be successful

The young adult stage is full of major changes in both physical and cognitive attributes. The body has finished fully developing and the thinking process is carried out in a more complex manner. It is during this development stage that the young adult can contemplate the views of others and put themselves in their place to gain a better understanding. Many key events in adulthood occur at relatively predictable times for most people in an age cohort. An age clock represents our internal sense of time for when major life events should occur. Physically, it is a time when one is at his or her healthiest. The brain is still increasing in size, although new neurons are no longer forming. One sense is also the keenest during this time of life. Full maturation has been reached, as well as full height. This is also the time when this age group learns to live comfortably in their own.

The young adult years are often referred to as the peak years. We may find that young adults experience excellent health, vigor, and physical functioning. Young adults have not yet been subjected to age-related physical deterioration, such as wrinkles, weakened body systems, and reduced lung and heart capacities. Their strength, coordination, reaction time, sensation, fine motor skills, and sexual response are at a maximum. Additionally, both young men and women enjoy the benefits of society's emphasis on youthfulness. They typically look and feel attractive and sexually appealing. It is quite clear that young men may have healthy skin, all or most of their hair, and well-defined muscles and young women may have soft and supple skin, a small waistline, and toned legs, thighs, and buttocks. Early in adulthood, neither gender has truly suffered from any double standard of aging, mainly, the misconception that aging men are distinguished, but aging women are over the hill. With good looks, great health, and plenty of energy, young adults dream and plan their future career and personality. Adults in their 20s and 30s set many goals that they intend to accomplish, from finishing graduate school, to getting married and raising children and to becoming a millionaire. Young adulthood is a time when nothing seems impossible; with the right attitude and enough persistence and energy, anything can be achieved. Some individuals begins habits that likely will produce health problems later in life such as overeating, overuse of alcohol, drugs and lack of exercise. Physical change may come in the form of weight gain for this age group. This is the time of settling into careers which can be sedentary, compared to the activities that are done in college and high school. For many, this is the first time in taking sole responsibility for providing nourishment. Many young adults move away from home. Food intake may now consist of fast food and frozen dinners, which can really rack up the pounds. Early adulthood is often the time during which people are most sexually active, and many plan to have children. Sexually transmitted diseases affect most of the young adults such as Chlamydia, gonorrhea, and syphilis. Aids may be least partly responsible for a shift to more caution sexual behavior.

Many young adults have developed the skill to reason logically and solve abstract problems. This is also the age when they are able to solve theoretical problems. It is here established that this age group scores higher than any other on the fluid intelligence section of an IQ test. Fluid intelligence is not only the ability to think abstractly, but to deal with novel situations. This is the age that awareness of consequences develops. Piaget argued that cognitive development reaches its highest level, their thinking becomes more complex.